RAINFED AGRICULTURE

nipa Co-Published
Meant For Sales Only in SAARC Countries

NEW INDIA PUBLISHING AGENCY
New Delhi-110 034

About the Authors

Dr. Jayadeva, H.M. obtained his B.Sc. (Agri.) in 1993 and M.Sc. (Agri.) in Agronomy in 1995 and Ph.D in 2007 from University of Agricultural Sciences, Bangalore. He started his professional career as Assistant Professor at UAS, Bangalore, later served as Associate Professor and Professor at UAS, Bangalore. Presently, he is serving as Professor and Head, Department of Agronomy, UAS, GKVK, Bangalore.

He has guided 09 masters degree students. He has published 98 peer reviewed research papers in National and International Journals. He developed Seven technologies to improve the yield of paddy and pulses. He received the best thesis award from Indian Society of Agronomy for his Ph.D. Thesis in 2007, ICAR best teacher award for the year 2014-15 and 02 of his M.Sc. student received the PPIC best thesis award in 2015 and 2016.

Dr. B.K. Ramachandrappa started his carrier as Research Assistant, RRS, Dharwad during 1982, later served as Assistant Professor, Associate Professor and Professor of Crop production at College of Agriculture, Bengaluru. He also served as Associate Director of Research for Zone-IV in his 33 years of service. He retired as Chief Scientist & Head, AICRP on Dryland Agriculture and University Head of Agronomy, UAS, GKVK. He is presently, serving as a ICAR-Emeritus Professor, Department of Agronomy, UAS, GKVK, Bengaluru. He was instrumental in implementing national and inter-national projects worth six crores in the University and has presented papers in the international seminars held at Thailand, Mexico and China. Further, under his able leadership the Dryland agriculture project, GKVK bagged 'National Best Center Award' and 'Most Responsive Center Award'.

He has developed water management practices for different crops as well as fertigation techniques. Sustainable dryland technologies viz., soil and water conservation, efficient water use, INM, cropping systems, dryland mechanization and contingency practices for aberrant weather conditions were standardized and demonstrated in the red soil regions of southern Karnataka under his leadership and has been recognized as an eminent Agronomist at the state and national level.

08 Ph.D. and 16 M.Sc. (Agri.) scholars received guidance under him in the University. He has over 250 National and international research publications to his credit besides, 18 books. He is also a recipient of Prestigious, Vasanth Rao Naik Award for Outstanding Research and Application in Dryland Farming Systems from ICAR, New Delhi for the year 2014; Dr. Kalayya Krishnamurthy National Award for the best Agricultural Research 2015-16; Gold Medal from the Indian Association of Soil and Water Conservationists' for the year 2015; Fellow of Indian Society of Oilseeds Research (2002-03) and National Environmental Science Academy (2015) are also the important other awards conferred on him.

RAINFED AGRICULTURE
Principles and Practices

Jayadeva, H.M.
Professor and Head
Department of Agronomy
University of Agricultural Sciences, GKVK
Bengaluru-560 065
Karnataka, India

B.K. Ramachandrappa
ICAR - Emeritus Professor
Department of Agronomy
University of Agricultural Sciences, GKVK
Bengaluru-560 065
Karnataka, India

Co-Published
Meant For Sales Only in SAARC Countries

NEW INDIA PUBLISHING AGENCY
New Delhi-110 034

Co-Published
Meant For Sales Only in SAARC Countries

NEW INDIA PUBLISHING AGENCY
101, Vikas Surya Plaza, CU Block, LSC Market
Pitam Pura, New Delhi – 110 034, India
Email: info@nipabooks.com
Web: www.nipabooks.com

For customer assistance, please contact
Phone: +91-11-27 34 17 17 Fax:+91-11-27 34 16 16
E-Mail: feedbacks@nipabooks.com

ISBN : 978-93-89907-03-2

Co-Published with CRC (CRC Press is an imprint of the Taylor & Francis Group, an informa business). Meant for Sales Only in India, Pakistan, Nepal, Sri Lanka and Bhutan. Other countries' distribution rights are vested with CRC.

Composed and Designed by NIPA.

Website : www.usabangalore.edu.in
E-mail : vc@usabangalore.edu.in
vcusab1964@gmail.com
srprasad1989@yahoo.co.in

✆ Off. : 080-23332442
: 080-23330153 (Extn. 265)
Mob. : 9449866900
Fax : 91-080-23330277

University of Agricultural Sciences Bangalore

DR. S. RAJENDRA PRASAD
Vice Chancellor

Gandhi Krishi Vignana Kendra
Bangalore-560065

Foreword

Rainfed areas in India are highly diverse, ranging from resource rich areas to resource constrained areas. Some of the resource rich areas are highly productive and have experienced widespread adoption of technology. However, most of the areas are resource constrained and dry areas. In resource constrained area, farming is a survival mechanism rather than a growth oriented activity. Rainfed crops prone to moisture stress due to variability of rainfall, delay in sowing, diversity in crop management practice and variation in the soil type. Most of the increase in agricultural output over the years has taken place under irrigated conditions. The opportunities for continued expansion of irrigated area are limited. So, Indian Policymakers are looking to rainfed agriculture to meet the rising demand for food projected over the next decades. In India, about 60 per cent of total sown area comes under rainfed lands. The rainfed areas contributes about 40 per cent of the country's food production and rural poor people depend on rainfed agriculture for their livelihood. Therefore, the development in the rainfed regions would be foremost importance in future.

Textbooks on Rainfed Agriculture-Principles and Practices suitable for undergraduate and post-graduate students are very few. This book on Rainfed Agriculture-Principles and Practices by Dr. Jayadeva, H. M. and Dr. B. K. Ramachandrappa is a timely and welcome contribution to fill gap in Rainfed Agriculture. The books is divided into thirteen chapters and covers comprehensively the content of Rainfed Agriculture offered for undergraduate and post-graduate students of Agricultural Universities in the country. The authors have considerable experience in teaching and research in Rainfed Agriculture and have presented the subject matter critically.

I hope this publication will serve as one of the model reference for students of undergraduate and postgraduate in strengthening the work on Rainfed Agriculture. I congratulate the authors for their efforts in bringing out this book covering all aspects of Rainfed Agriculture.

S. Rajendra Prasad
Vice Chancellor

Preface

The rainfed agriculture distributed approximately ¾th of global crop lands, is expected to respond to climate change and human population growth and these responses may be especially pronounced in water limited areas. Because the environmental conditions that support rainfed agriculture are determined by climate, weather, and soil conditions that affect overall and transient water availability, Indian Agriculture is being described even today as a gamble of the monsoon and every year fate of farmers oscillate with the behaviour of monsoon. Different types of droughts of varied intensities experienced at a different growth stages *viz*., early, mid and end season drought are the common feature of the rainfed ecosystem.

In this publication entitled "Rainfed Agriculture-Principles and Practices" an earnest attempt has been made to include various issues related to Rainfed Agriculture *viz*. soil and water conservation, drought management strategies, rain water harvesting, crops and cropping system for dryland agriculture, conservation agriculture, climate-smart agriculture and watershed management as per the recommendation of fifth Dean's Committee of Indian Council of Agricultural Research has been presented. The subject matter is both narrative and critical. Illustrations have been added to make the subject matter more clear.

Authors are highly indebted to the contributions of the various reviews through their books, monographs, periodicals and reports, especially of AICRPDA, Bangalore centre reports from which liberally research data has been drawn to demonstrate the worthiness of dryland technology in this textbook. The authors are pleased to express their gratitude for encouragement to all the colleagues and students of department of agronomy.

The authors will be grateful to the teachers and students, if they point out the error and send suggestions and constructive criticism for future revision and improvement of the book.

The authors would like to place on record their sincere gratitude to Dr. S. Rajendra Prasad, Vice Chancellor, University of Agricultural Science Bangalore for writing the foreword to this book.

Authors

Contents

Chapter 1

Introduction and History of Rainfed Agriculture

The term **Rainfed agriculture** is used to describe **farming** practices that rely on rainfall for water. It provides much of the food consumed by poor communities in developing countries. In India, about 52% of total net sown area comes under rainfed lands. Rainfed crops account for 48 per cent area under food crops and 68 per cent under non-food crops. India ranks first among the rainfed agricultural countries of the world in terms of both extent and value of produce. Due to population pressure on agricultural lands, the poverty is concentrated in rainfed regions. The climate in India's rainfed regions is characterized by complex climatic deficiencies, manifested as water scarcity for rainfed crop production. The climate is largely semi-arid and dry sub-humid with a short (occasionally intense) wet season followed by long dry season. Rainfall is highly unreliable, both in time and space, with strong risks of dry spells at critical growth stages even during good rainfall years. The fluctuations are due to numerous factors affecting the monsoonal climate including the atmospheric circulation and strong links to El-Nino Southern Oscillation (ENSO) phenomenon in the Pacific Ocean.

Rainfed areas in India are highly diverse, ranging from resource rich areas to resource-constrained areas. Some of the resource rich areas are highly productive and have experienced widespread adoption of technology. However, most of the areas are resource constrained and dry areas. In the resource constrained and dry areas, the farming is a survival mechanism rather than a growth oriented activity. Rainfed agriculture is practiced under a wide variety of soil type, agro-climatic and rainfall conditions ranging from 400 mm to 1600 mm per annum. Rainfed Crops are prone to breaks in the monsoon during the crop growth due to water stress. This water stress may be due to variability of rainfall, delay in sowing, diversity in crop management practice and variability of the soil type. The prolonged breaks can result in partial or complete failure of the crops.

In the past, the Rainfed farming system was mainly dependent upon the locally available inputs (seeds, manures, animal draught) and used to grow a number of crops, which were able to withstand drought-like situation. However, in recent times, the cropping systems have changed and currently the farmers in these rainfed areas have limited options. Many of the farmers in these regions started cultivating high value crops which requires intensive use of costly inputs (chemical fertilizers/pesticides, hybrid seeds, life saving irrigation, farm energy *etc*.,) and find it difficult to manage the resources on their own. To reduce their vulnerability to rains, farmers in some areas grow crops such as jowar, bajra and pulses. These crops are low-yielding, but less affected by variations in rainfall. This saved the farmers from the risky nature of farming in rainfed and dry areas.

Dry farming or dry land agriculture has not received sufficient attention as they deserve and green revolution had certainly left the dry farming area in backlog. Their contribution was not much as compared to irrigated areas, may be due to continued low productivity, inhabitancy by poorest segments of the country, crop failures, highly erratic and unpredictable rainfall pattern, intermittent dry spells during crop growth, neglected soil and crop management practices and so on. There is considerable yield gap in various dryland crops compared to the potential yield indicating the need for concerted efforts through comprehensive approaches to tap the yield potential (Table 1).

Table 1: Yield levels of important dry land crops in farmers fields in comparison with their potential yield (kg ha^{-1}) in the agro-climatic zones of south Karnataka

Crop	PY (potential yield)	Farmer yield (Zone-4)	Yield gap (%)	Farmer yield (Zone-5)	Yield gap (%)	Farmer yield (Zone- 6)	Yield gap (%)
Fingermillet	4000	2473	1527 (38.2)	2939	1061 (26.5)	2151	1849 (46.2)
Maize	5900	3510	2390 (40.5)	3510	2390 (40.5)	3414	2486 (42.1)
Bajra	1400	546	854 (61.0)	595	805 (57.5)	540	860 (61.4)
Redgram	1600	426	1174 (73.4)	453	1147 (71.7)	328	1272 (79.5)
Horsegram	1000	566	434 (43.4)	427	573 (57.3)	421	579 (57.9)
Black gram	750	472	278 (37.1)	429	321 (42.8)	415	335 (44.7)
Greengram	1200	461	739 (61.6)	336	864 (72.0)	324	876 (73.0)
Cowpea	1250	566	684 (54.7)	515	735 (58.8)	498	752 (60.2)
Groundnut	1000	908	92 (9.2)	1220	–220 (–22.0)	1017	17 (1.7)
Sunflower	1900	548	1352 (71.2)	442	1458 (76.7)	514	1386 (72.9)
Sesame	500	595	95 (19.0)	437	63 (12.6)	444	56 (11.2)

(Figures in parenthesis indicates per cent difference in yield gap)

Table 2: Crop wise yield gaps (kg ha^{-1}) in the operational research project of AICRPDA, Bengaluru.

Crop	Experimental yield (kg ha^{-1})	Attainable yield (Farmers yield with improved technology) (kg ha^{-1})	Farmers Yield (kg ha^{-1})	Best bet practices for up scaling and bridging yield gap*
Finger millet	4000	3644	1580	Intercropping with Pigeonpea
Maize	5000	4000	2500	Crop rotation, intercropping with pigeonpea and application of micronutrients
Pigeonpea	1600	1452	800	Wider spacing
Horsegram	1000	800	500	Line sowing and weed control
Cowpea	1250	1000	700	Application of micronutrients
Groundnut	1500	650	400	Application of secondary and micronutrients and intercropping with Pigeonpea
Sunflower	1800	1200	800	Application of secondary and micronutrients and thinning

*15 to 20% increase in yield by adopting *in-situ* mositure conservation practices in dryland areas.

Source : Ramachandrappa *et al.*, 2014

Rainfed Agriculture

The term rainfed agriculture or dryland agriculture may be defined as growing of crops entirely under rain water. Dryland agriculture is classified into three categories depending upon the amount of rainfall received *viz.*, dry farming, dryland farming and rainfed farming. The distinction exists between these three, depending on annual rainfall and other parameters (Table 3).

Dryfarming: Cultivation of crops in regions with annual rainfall less than 750 mm. Crop failure is most common due to prolonged dry spells during the crop period. These are arid regions with a growing season (period of adequate soil moisture) less than 75 days. Moisture conservation practices are most important in these regions.

Dryland farming: Cultivation of crops in regions with annual rainfall more than 750 mm. In spite of prolonged dry spells crop failure is relatively less frequent. Dryland farming areas are grouped under semiarid tracts with a growing period between 75 and 120 days. Moisture conservation practices are necessary for crop production. However, adequate drainage is required especially for Vertisols/black soils.

Rainfed farming: Crop production in regions with annual rainfall more than 1150 mm. Crops are not subjected to soil moisture stress during the crop period. Emphasis is often on disposal of excess water. These are humid regions with growing period more than 120 days.

Table 3: Distinction between dry farming, dryland farming and rainfed farming

S. No.	Parameter	Dry Farming	Dry land farming	Rainfed farming
1.	Rainfall	< 750 mm	750–1150 mm	> 1150 mm
2.	Region	Arid	Semi-Arid	Semi arid-Humid
3.	Risk	Frequent	Less frequent	Safe
4.	Conservation Practices	Conservation of soil & Water	Conservation of soil & Water + Drainage in Black soils	Conservation & disposal of Excess water
5.	Cropping	Mono cropping	Mono Cropping & Inter Cropping	Double cropping
6.	Growing period	< 75 Days	75 – 120 Days	> 120 Days
7.	Erosion	Wind	Wind & Water	Water
8.	Management	Extensive	Moderate	Intensive
9.	Yield	Less stable	Moderately stable	Sustainable

Source: Anon., 2019

United Nations Economic and Social Commission for Asia and the Pacific distinguished dry land agriculture into two categories: dryland and rainfed farming (Table 4).

Table 4: Distinction between dryland farming and Rainfed farming

Constituent	Dryland farming	Rainfed farming
Rainfall (mm)	< 800	> 800
Moisture availability to the crop	Shortage	Enough
Growing season (days)	< 200	> 200
Growing regions	Arid and semiarid as well as uplands of sub-humid and humid region	Humid and sub-humid region
Cropping system	Single crop or intercropping	Intercropping or double cropping
Constraints	Wind and water erosion	Water erosion

Source: Anon., 2017

Scope and Importance of Rainfed Agriculture

1. Out of 141.427 m.ha net sown area in India, 51.84% (73.327 m.ha) depends on rainfall (Table 5).

Table 5: Pattern of land utilization in India during 2013–14 (Area: in million hectares)

Total Geographical Area	Net sown area	Current fallows	Total cultivated	Net irrigated	Net rainfed
328.726	141.427	14.154	155.581	68.10	73.327

Source: Directorate of Economics and Statistics, Ministry of Agriculture, 2017.

2. Rainfed area contributes significantly both in area and production in India

Table 6: Distinction of crops under Rainfed Agriculture

Area Contribution		Production Contribution	
Sorghum,Groundnut, Pulses	> 90%	Total Food grains	45%
Maize, Chick Pea	82–85%	Pulses	90%
Rape seed/Mustard	65.8%	Oilseeds	75%
Cotton	78%	Cotton	70%
Rice	61.7%		
Barley	44%		
Wheat	35%		

3. About 70% of rural population depends on success or failure of crops and in turn rainfall for their livelyhood.
4. Although, India is blessed with average annual rainfall of about 1200 mm, slightly above the global mean of 990 mm, the fate of dryland crops oscillates with the quantity, onset, and progress, spatial and temporal distribution of monsoon rains. Of the mean annual rainfall, 30 per cent of the country gets less than 750 mm and 40 per cent between 750 and 1250 mm. Only 20 per cent area is blessed with rainfall between 1250 and 2000 mm, leaving about 10 per cent area with annual rainfall over 2000 mm.
5. Much of the increase in food production in the recent past is estimated to be mainly due to irrigated areas. According to experts, even when the ultimate irrigation potential is reached, 55 per cent of the net sown area will be still, rainfed. Therefore, it is inevitable that in future the additional food has to come mainly from the drylands.

6. At present, 3 ha of dryland crop produce cereal grain equivalent to that produced in one ha irrigated crop. Hence, there is a great scope to increase productivity under rainfed situation.

	Rainfed	Irrigated
Average productivity (t ha^{-1})	0.4–0.7	1.5–2.0

7. The per capita land availability, which was 0.28 ha in 1990, has declined to 0.19 ha with limited scope for increasing the area under plough. Only option left is to increase the productivity with the modern technology and inputs in drylands.
8. The productivity has already showed a plateau in irrigated agriculture due to problems related to nutrient exhaustion, salinity build up and depleting water table. Hence, the challenges are to make grey areas (drylands) as green.
9. Dryland offers good scope for development of Agroforestry, Social forestry, Horti-Silvi-pasture and other similar systems which will not only supply food, fuel and fodder, but forms a suitable vegetative cover for ecological maintenance.
10. Dry/Rainfed land offers vast scope for increasing cropping intensity and employment generation. The cropping intensity in rainfed areas is 110–115% against 150–300% in irrigated areas.

History of Rainfed Agriculture

Rainfed agriculture dates back to the origin of Agriculture itself. The systematic study of constraints in dry farming was initiated during **1862** when USA introduced Morill Bill Act. In this act, emphasis has been made to impart agricultural education through the establishment of Agricultural College. Due to severe scarcity of food, a famine commission was appointed in **1880** to provide suitable measure to overcome famine and started protective irrigation projects.

Meanwhile, H.W. Campbell, a progressive farmer started cultivating wheat on large scale (300 ha) and harvested a bumper crop (12000 Bushels). However, the crop was totally failed during **1882** because of drought and he started investigating reasons for such failure. He tried to trace back the historical literatures and focused his attention on the suggestions of the Father of Tillage, Jethrotull. In **1894,** 1st Dry Farming Experiment station was started at Cheyennue Well in Collarado State of America.

Campbell published a book 'Soil Culture and Farm Journal Manual' in **1880**. He emphasized on the importance of summer tillage in harvesting early rains during **1898**.

The historical premise for dry farming was explained by a Utah proponent Dr. John A. Widstoe pioneer of dryland farming research and President of State Agricultural College, Utah, USA in his book *Dry-Farming, A System of Agriculture for Countries Under Low Rainfall*, defined dry farming as "the profitable production of useful crops without irrigation on lands that receive rainfall of less than 500 mm annually". Widstoe summarized the practices for dry farming in his one more book "Gospels of Dry farming" as follows:

a. Deep ploughing and Fall ploughing, b. Early sowing of crops c. Raising drought resistant crops d. Fallowing alternate years under very low rainfall areas (<15 inch) and once in three years under low rainfall situation (15–20 inch)

1901: USA passed Arid Farm Bill and emphasized to start dry farming research stations.

1904: Campbell summarized his findings and recognized as 'Campbell System of Dry Farming'. Practices advocated by Campbell are

(a) Disc ploughing immediately after every harvest
(b) Shallow light tillage operations after discing
(c) Passing sub - soil packer
(d) Smooth harrowing
(e) Thin sowing using seed drill

1907: A forum of progressive farmers and scientists were created to exchange ideas in the name of "**1st Dry Farming Congress**". Subsequently, 2nd & 3rd Dry Farming Congress were held during 1908, 1909, respectively.

In India

Dryland agriculture in India has been practiced since time immemorial. Unlike USA, Australia, Canada where dryland agriculture is mechanized, Farmers in India developed innovative methods to grow crops under dryland conditions.

Government of India, as per Famine and Royal Commission recommendations established IARI and ICAR. Later in mid 50's, a renewed efforts were made to conserve natural resources to attain food security through establishment of Soil Conservation Research Centers at 8 locations *viz.*, Dehradun, Chandigarh, Ooty, Bellary, Kota, Vasad (Guj), Agra and Hyderbad.

The chronology of events in dryland agricultural research in India is :

1920: Scarcity tract development given importance by the Royal Commission on Agriculture

1923: Establishment of Dryland Research Station at Manjiri (Pune) by Tamhane

1933: Research Station established at Bijapur and Solapur

1934: Research Station established at Hagari and Raichur

1935: Research Station established at Rohtak (Punjab)

1942: Bombay Land Development act passed

1944: Monograph on dry farming in India by N. V. Kanitkar (Bombay, Hyderbad, Madras Dry Farming Practices)

1953: Established Central Soil Conservation Board

1954: Established Central Soil Conservation Centres, Bellary

1970: Establishment of the All India Co-ordinated Research Project for Dryland Agriculture (AICRPDA) as a collaborative project between the Indian Council of Agricultural Research (ICAR) and Canadian International Development Agency (CIDA). The AICRPDA was initiated at 23 centers selected on agro-climatic basis. These centers were under the State Agricultural Universities or ICAR Research Institutes

1972: ICRISAT was established under CGIAR at Patancheru, Hyderbad to take care of research needs for semi-arid tropics

1974: Establishment of five Operational Research Centers were also initiated to test the strength and weakness of the research findings on farmers fields. These centres are at Hyderabad, Bangalore, Ranchi, Hoshiarpur and Indore. During the VIth Five Year Plan, four more were added at Anantapur, Solapur, Hisar and Udaipur.

The primary mandate for the project was to improve and stabilize crop production in dry lands of India through the following objectives

- To evolve simple and easily implementable crop production practices to increase production at least by 100 per cent,
- To stabilize the production over years by providing alternate crop production technologies to match weather aberrations, and
- To optimize the use of natural resources *viz.,* land and water (rainfall).

1983: Started 47 model watersheds under ICAR

1984: Initiation of World Bank assisted "Watershed Development Programmes in four states". Establishment of Dryland Development Board in Karnataka

1985: Birth of Central Research Institute for Dryland Agriculture at Hyderbad

1990–91: Launching of NWDPRA (National Watershed Development Project for Rainfed Area) started in 25 states and 2 Union territories.

Chapter 2

Constraints of Rainfed Areas

With renewed interest in rainfed agriculture during 1970s, both the national (AICRPDA and CRIDA) and international institutions (ICRISAT) have studied several aspects of rainwater management. Recommendations have been made after their through testing both at the research Stations and on farmer's fields. Initially these technologies were accepted by the farmers because of the pursuation by the Scientists and the incentives, but many of them are forgotten once the support is withdrawn. It is now time for us to examine whether the technologies themselves were weak or certain components were missing malfunctioning of the software may be due to a bug right from the beginning or it entered into the system at a later stage. It is also possible that we neglected certain critical inputs which became major constraints making the system defunct. Whatever the reason it should be identified and set right if the technology has to continue as tool for tackling the challenges of rainfed agriculture.

Main constraints of dry land agriculture to enhance the productivity are classified as follows:

I. Ecological constraints
II. Edaphic constraints
III. Socio-economic constraints
IV. Technological constraints
V. Transfer of technology constraints

I. Ecological Constraints

1. Rainfall

(i) Low and erratic rainfall distribution

In dryland area, deficiency and uncertainty in rainfall of high intensity causes excessive soil erosion which leaves the soil unproductive. Owing to erratic behavior and distribution of rainfall, agriculture is risky, farmers lack resources, tools become inefficient ultimately productivity is low.

(ii) Variable Rainfall

Annual rainfall varies greatly from year to year and its coefficient of variation. Generally, higher the rainfall less is the coefficient of variation. In other words, crop failures due to uncertain rains are more frequent in regions with lesser rainfall.

(iii) Intensity and distribution

In general, more than 50 per cent of total rainfall is usually received in 3 to 5 rainy days. Such intensive rainfall results in substantial loss of water due to surface runoff. This process also accelerated soil erosion. Distribution of rainfall during the crop growing season is more important than total rainfall in dryland agriculture.

(iv) Late onset and early cessation of rains

Due to late onset of monsoon, sowing of crop is delayed resulting in poor yields. Sometimes the rain may cease very early in the season exposing the crop to drought during flowering and maturity stages which reduces the crop yields considerably.

(v) Prolonged dry spells during crop period

Long breaks in the rainy season are an important feature of Indian monsoon. These intervening dry spells when prolonged during crop period reduces crop growth and yield and when unduly prolonged crops fail.

II. Edaphic Constraints

1. **Alfisols:** These are commonly referred as red soils. Problems relating to crop production are poor crop stand due to crusting and rapid drying of surface soil. Poor crop growth due to unreliable soil moisture supply, low moisture storage capacity due to shallow depth and drought spells during crop season. These soils are low in organic matter, poor in nutrient status with respect to N, P, S and Ca. Compact subsoil layer (argillic horizon). Land degradation due to soil erosion and crusting.

These are commonly referred as red soils. Red soils have usually shallow depth and low moisture retention capacity. The water holding capacity of these soils varies between 10 and 15 cm per metre depth and this is not sufficient to carry the crops for more than a month after the cessation of rains. Therefore, only *kharif* crop is possible. The constraints relating to crop production are

- Crusting on drying is a serious problem in seedling emergence and establishment of a good crop stand. When heavy rains occur soon after sowing followed by a dry spell, there can be complete failure of germination. Under uncultivated conditions, this leads to high runoff in the order of about 25 per cent.

- Poor crop growth due to unreliable soil moisture supply, low moisture storage capacity due to shallow depth and drought spells during crop season.
- Red soils are normally poor in NPK and very low in organic matter status. At places, they are also deficient in Ca, S, Zn and B. One of the ways to make them productive is to improve the organic matter status.
- The soils have low CEC (10–15 meq/100 g) which results in low retention of exchangeable nutrients. In fairly high rainfall areas, the nutrients can be easily leached.
- The soils are highly erodible. As per the estimation available, the annual soil loss in red soil is 4 to 10 t ha^{-1}. When such erosion is not controlled in the beginning, it can lead to formation of gullies. It is estimated that about 15 per cent of the area of red soils is under gullies.

2. **Vertisols:** These soils commonly called as black soils are characterized by high clay content (30–70%), important constraints for crop production are:
   - Physical constraints such as narrow range of soil water content for tillage, tendency to become waterlogged and poor tractability.
   - Low infiltration and highly susceptible to soil erosion due to predominance of silt and clay
   - Poor drainability and susceptible to water logging
   - Land degradation due to soil erosion and salt accumulation, especially in low lying area.
3. **Inceptisols and Entisols:** These are commonly termed as alluvial soils. They have low water holding capacity and low nutrient holding capacity. Management of these soils for crop production is relatively easy compared to red and black soils. Soil erosion is however a problem leading to degradation.
4. **Sub mountain soils:** Sub mountain soils are medium in texture and depth is medium to deep as well as more clay content. Moisture retention capacity is high (300 mm m^{-1} profile depth). These soils are poor in nitrogen. Phosphorus may be limiting in high production system. Due to high water holding capacity double cropping is possible in these soils.
5. **Sierozems:** Sierozems are extremely light soils, effective depth being influenced by the $CaCO_3$ concentration in soil profile. Its moisture holding capacity is low (150 mm m^{-1} soil depth). Sierozemic soils are low in nitrogen and sometimes inadequate in phosphorous. Subsoil salinity is common. These soils are mostly monsoon cropped, except in deep sandy loams where double cropping is also possible. Crusting is very frequent.

III. Socio-economic Constraints

The socio-economic status of dryland farmers, generally, will not permit them in adopting recommended dryland technology. Major socio-economic constraints are:

1. Lack of capital and credit at proper time make the farmers hesitate to adopt the recommended technology.
2. Most of the dryland farmers are resource poor. The farmers engage their family labour force for most of the farm activities.
3. The organizational structure for input supply is very poor in dryland areas.
4. The socio-economic survey conducted in several places have indicated that more than 50 per cent of the farmers do not have their own bullock pairs and there is a great constraint in timely land preparation and sowing.
5. The land holdings are very small. About 75 per cent of the holdings in the state are small and marginal, having a size of 1.6 ha. Such small holdings are not economical under dryland situations and pose problems in implementing land development programme.
6. Due to poor economic status of the farmers, they do not have investment capacity and risk bearing status essential in dryland conditions.
7. Lack of support price for dryland crops and marketing problems and predominant role of middle man.
8. Inadequate availability of agriculture inputs during the season and for contingent measures due to weather aberrations.

IV. Technological Constraints

The technologies developed by universities through research stations for different dry farming areas are highly effective. Adoption at farmers field is very less, because of the following reasons

1. Lack of high yielding varieties for adverse climatic conditions as compared to local varieties.
2. Technologies to utilize the improved machineries/implements under dry lands by small and marginal farmers leading to untimeliness of field operations and crop failures/reduce yields.
3. Adoption of poor pest and disease management, due to non availability of water in dry land areas
4. Adoption of alternate land use systems *viz.,* agro forestry, alley cropping, dry land horticulture in dry lands is very less

5. Extension activities is inadequate leading to poor spread of technologies *viz.*,
   - Soil and moisture conservation
   - Crops and crop varieties
   - Climate resilience in agriculture technology
   - Use of inputs such as herbicides, fertilizers, pesticides *etc.*
6. The main difficulty in estimating the extent of runoff is the non availability of the data on daily rainfall and rainfall intensity for each watershed/ catchment. In this direction, the efforts of the KSNDMC in Karnataka and similar organizations elsewhere to install telemetric rain gauges at panchayat land needs to be recommended. Such data could be used for estimation of runoff in each watershed and storage structures planned accordingly.
7. Non availability of low cost and efficient water lifting devices: It is a common observation in watersheds that stored water in ponds is often not used for irrigations. It is not that farmers do not know the benefits of protective irrigation, but the reason is non availability of low cost and efficient water lifting devices. This single limitation has made the farm ponds defunct in several watersheds. Though some executive argue that the ponds are constructed solely for recharge of groundwater and least importance to be given to protective irrigation, this cannot satisfy small farmers as they have to lose a portion of their fertile land for pond digging and some more on its annual maintenance. If farm ponds have to be attractive to farmers, suitable lifting devices has to be identified and made available at affordable prices. Krishi Bhagya flagship programme of the government of Karnataka in dry land areas is addressing all these issues as a comprehensive programme for enhancing the overall productivity.

V. Transfer of Technology Constraints

1. Though *in-situ* water conservation practices are simple to follow they are the ones least appreciated by several executives in watershed development department. The main argument is that there is not much for the departments staff to do and once the principles are explained, the farmers should adopt it by themselves. Several practices like ‘fall ploughing’ and contour farming, though appreciated by the farmers are not easily adopted as they lack the required power to take up cultivation on time.
2. Survey conducted by the team of operational Research Scientists in Bangalore showed that more than 50% of the farmers do not own even a bullock pair. Custom hire services or societies owning required

machinery can solve this problem. Further, what is required is, affordable power equipment, which small farmers can buy and maintain individually. Another important finding, which is extremely useful is organic mulching. Yet the technology is not a practice by the farmers. The reason is ‘sufficient mulching material is not available’. But is it a statement made after making honest efforts to collect the weeds removed from cropped fields. Green manure plants like gliricidia crop residues and leaf litter from trees on the farm bunds and roadsides? This may not be a case for developing a new technology but an example to plead for greater interaction with farmers. Large number of demonstration and Farmer Field School (FFS) approach will be essential to convince the farmers and follow it as a routine.

3. Inter terrace management practices are not necessarily for conservation of rainwater, at times they can be for smooth disposal of surplus water. When the time and intensity of rainfall are uncertain, a dry land farmer has to be ready with both the contingencies of surplus and deficiency.
4. Ridge and furrow method of cultivation on a gradient conserves water when intensity of rainfall is small but the same furrow can act as a disposal system when intensities are higher. A paired row of Redgram with deep furrow in between recommended in red soil areas of Karnataka in Ragi and Groundnut acts as a dual purpose treatment and meets this requirement.
5. In spite of the availability of viable technologies for rainfed areas, most of them have not been adopted. The reason appears to be lack of convincing demonstrations. Such demonstration should focus on an integration of several aspects of water conservation and its use and should be laid out at the research stations, KVKs and on farmers fields.

Chapter 3

Dry Land Soils

Soil is the top layer of the earth surface in which plants can grow, consisting of rock and mineral particles mixed with decayed organic matter and having the capability of retaining water. Soil is a mixture of organic matter, minerals, gases, liquids, and organisms that together support life. Earth's body of soil, called the pedosphere, has four important functions:

- as a medium for plant growth
- as a means of water storage, supply and purification
- as a modifier of Earth's atmosphere
- as a habitat for organisms

All of these functions, in their turn, modify the soil. The pedosphere interfaces with the lithosphere, the hydrosphere, the atmosphere, and the biosphere. Soil consists of a solid phase of minerals and organic matter (the soil matrix), as well as a porous phase that holds gases (the soil atmosphere) and water (the soil solution). Accordingly, soil scientists can envisage soils as a three-state system of solids, liquids, and gases.

The mineral components of soil are sand, silt and clay, and their relative proportions determine a soil's texture. Properties that are influenced by soil texture include porosity, permeability, infiltration, shrink–swell rate, water-holding capacity, and susceptibility to erosion. In the illustrated USDA textural classification triangle, the only soil in which neither sand, silt nor clay predominates is called loam. While even pure sand, silt or clay may be considered a soil, from the perspective of conventional agriculture a loam soil with a small amount of organic material is considered "ideal". The mineral constituents of a loam soil might be 40% sand, 40% silt and the balance 20% clay by weight. Soil texture affects soil behaviour, in particular, its retention capacity for nutrients (*e.g.*, cation exchange capacity) and water.

The **Indian Council of Agricultural Research** set up an All India Soil Survey Committee in 1953 which divided the Indian soils into eight major groups. For ease of understanding and comparative study, this post about the 8 soil types is prepared in tabular form, highlighting the key differences (**Table 7**).

Table 7: Different soils groups of India

Types of Soils	States where found	Rich in	Lacks in	Crops grown
Alluvial	Mainly found in the plains of Gujarat, Punjab, Haryana, UP, Bihar, Jharkhand *etc.*	Potash and Lime	Nitrogen and Phosphorous	Large variety of rabi and kharif crops such as wheat, rice, sugarcane, cotton, jute *etc.*
Black (Regur soil)	Deccan plateau-Maharashtra, Madhya Pradesh, Gujarat, Andhra Pradesh,Tamil Nadu, Valleys of Krishna and Godavari	Lime, Iron, Magnesia and Alumina, Potash	Phosphorous, Nitrogen and organic matter	Cotton, sugarcane, jowar, tobacco, wheat, rice etc.
Red	Eastern and southern part of the deccan plateau, Orissa, Chattisgarh and southern parts of the middle Ganga plain	Iron and Potash	Nitrogen, Phosphorous and humus.	Wheat, rice, cotton, sugarcane and pulses
Laterite	Karnataka, Kerala, Tamilnadu, Madhya Pradesh, Assam and Orissa hills	Iron oxide and potash	Organic matter, Nitrogen, Phosphate and Calcium	Cashewnuts, tea, coffee, rubber
Arid and Desert	Western Rajastan, north Gujarat and southern Punjab	Soluble salts, phosphate	Humus, Nitrogen	Only drought resistant and salt tolerant crops such as barley, rape, cotton, millets maize and pulses
Saline and Alkaline	Western Gujarat, deltas of eastern coast, Sunderban areas of West Bengal, Punjab and Haryana	Sodium, Potassium, Magnesium	Nitrogen and Calcium	Unfit for agriculture

Source: www.clearias.com/different-soil-types-india.

Characteristics of Dryland soils

Dryland soils vary in physicochemical as well as biological properties of the soil. These properties decide the infiltration rate, moisture-holding capacity of the soil, drainage pattern, aeration, compaction, soil erosion, organic matter decomposition, nutrient availability and crop growth. In India, drylands occur in almost all the regions of the country and the major soil types in these dry regions are red soils (*Alfisols*), desert soils (*Aridisols*) and black soils

(*Inceptisols* and *Vertisols*). The characteristic features of the major soil groups and the prospects for raising crops in these soils under dryland conditions are described here under:

Alluvial soil

- Mostly available soil in India (about 43%) which covers an area of 143 sq. km.
- Widespread in northern plains and river valleys.
- In peninsular-India, they are mostly found in deltas and estuaries.
- Humus, lime and organic matters are present.
- Highly fertile.
- Indus-Ganga-Brahmaputhra plain, Narmada-Tapi plain *etc.,* are examples.
- They are depositional soil-transported and deposited by rivers, streams *etc.*
- Sand content decreases from west to east of the country.
- New alluvium is termed as Khadar and old alluvium is termed as Bhangar.
- Colour: Light Grey to Ash Grey.
- Texture: Sandy to silty loam or clay.
- Rich in: potash
- Poor in: phosphorous.
- Wheat, rice, maize, sugarcane, pulses, oilseed *etc.,* are cultivated mainly.

Red soil

- Seen mainly in low rainfall area.
- Also known as Omnibus group.
- Porous, friable structure.
- Absence of lime, kankar (impure calcium carbonate).
- Deficient in: lime, phosphate, manganese, nitrogen, humus and potash.
- Colour: Red because of Ferric oxide. The lower layer is reddish yellow or yellow.
- Texture: Sandy to clay and loamy.
- Wheat, cotton, pulses, tobacco, oilseeds, potato *etc.,* are cultivated.

Black soil/regur soil

- Regur means cotton-best soil for cotton cultivation.
- Most of the Deccan is occupied by Black soil.
- Mature soil.
- High water retaining capacity.

- Swells and will become sticky when wet and shrink when dried.
- Self-ploughing is a characteristic of the black soil as it develops wide cracks when dried.
- Rich in: Iron, lime, calcium, potassium, aluminum and magnesium.
- Deficient in: Nitrogen, Phosphorous and organic matter.
- Colour: Deep black to light black.
- Texture: Clayey.

Laterite soil

- Name from Latin word 'Later' which means Brick.
- Become so soft when wet and so hard when dried.
- In the areas of high temperature and high rainfall.
- Formed as a result of high leaching.
- Lime and silica will be leached away from the soil.
- Organic matters of the soil will be removed fast by the bacteria as it is high temperature and humus will be taken quickly by the trees and other plants. Thus, humus content is low.
- Rich in: Iron and Aluminum
- Deficient in: Nitrogen, Potash, Potassium, Lime, Humus
- Colour: Red colour due to iron oxide.
- Rice, Ragi, Sugarcane and Cashew nuts are cultivated mainly.

Desert/arid soil

- Seen under Arid and Semi-Arid conditions.
- Deposited mainly by wind activities.
- High salt content.
- Lack of moisture and Humus.
- Kankar or Impure Calcium carbonate content is high which restricts the infiltration of water.
- Nitrogen is insufficient and Phosphate is normal.
- Texture: Sandy
- Colour: Red to Brown.

Peaty/marshy soil

- Areas of heavy rainfall and high humidity.
- Growth of vegetation is very less.

- A large quantity of dead organic matter/humus which makes the soil alkaline. Heavy soil with black colour.

Forest soil
- Regions of high rainfall.
- Humus content is less and thus the soil is acidic.

Mountain soil
- In the mountain regions of the country.
- Immature soil with low humus and acidic.

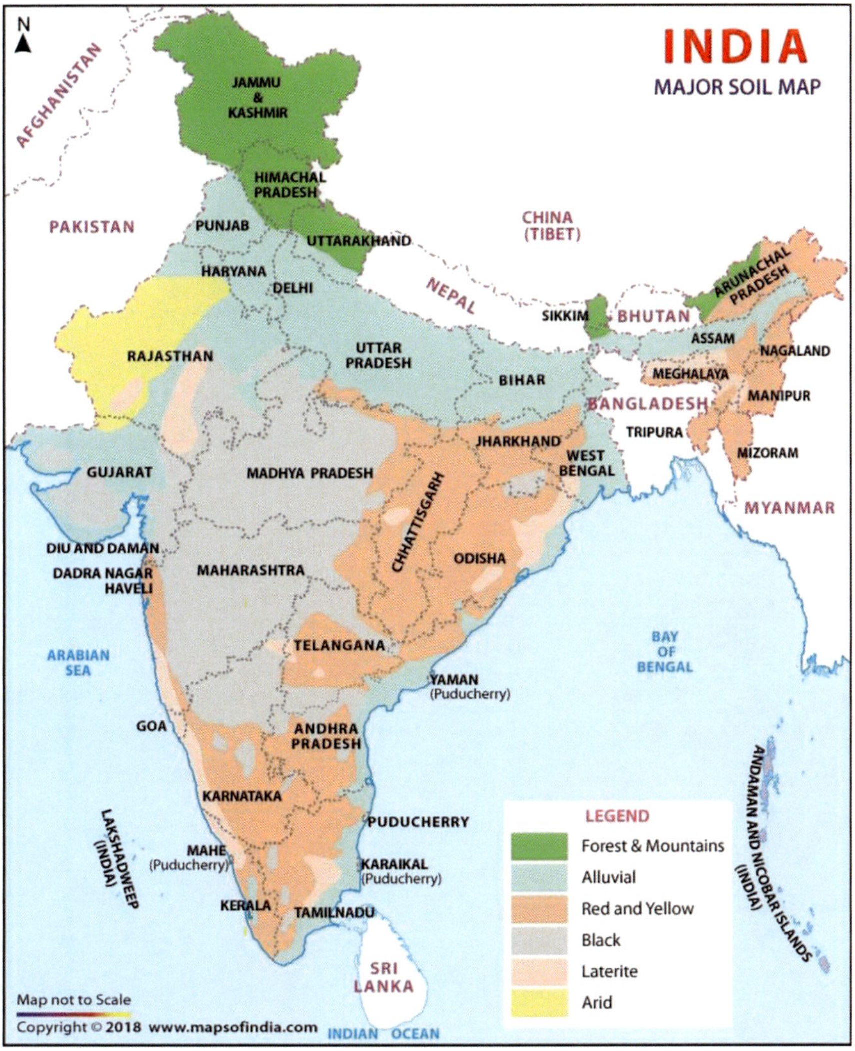

Deficiency of Nutrients Map in Drylands

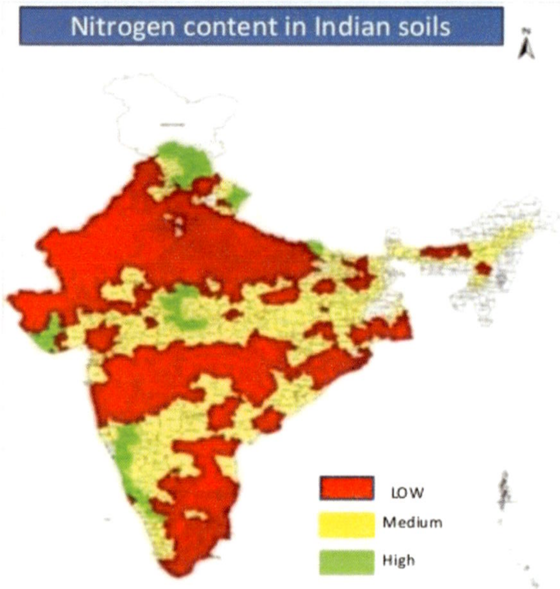

Source: IISS, Bhopal

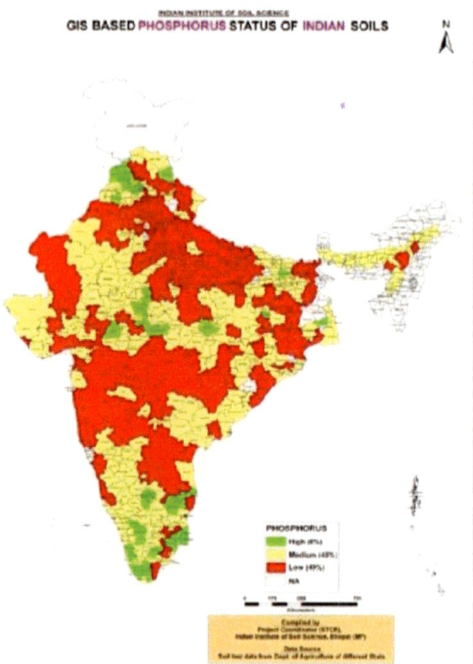

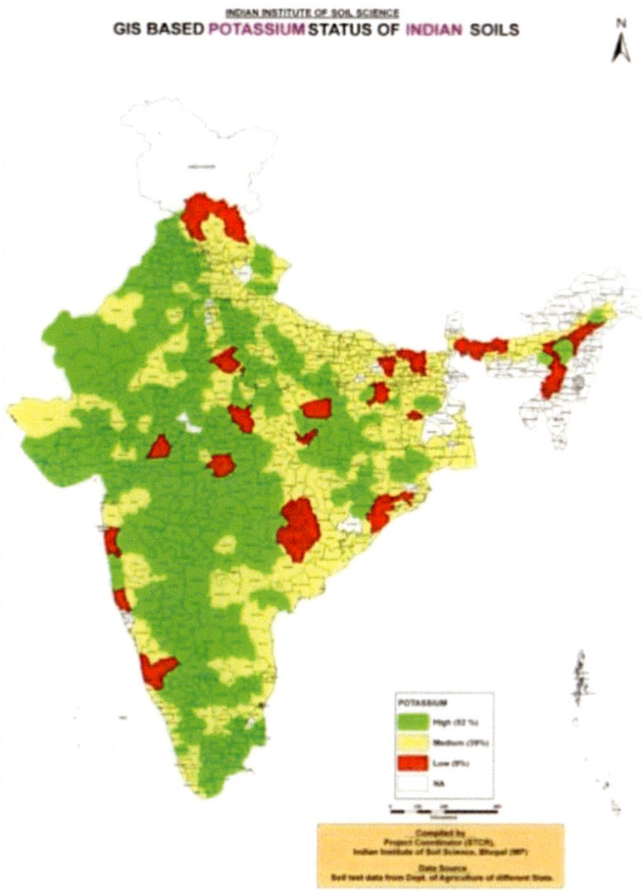
INDIAN INSTITUTE OF SOIL SCIENCE
GIS BASED POTASSIUM STATUS OF INDIAN SOILS
N
POTASSIUM
NA

Chapter 4

Meteorology

Meteorology is defined as the branch of science concerned with the processes and phenomena of the atmosphere. It is also defined as the climate and weather of a region. The word meteorology is derived from two Greek words: *Meteor* and *logus*. *Meteor* means event occurring above the earth surface and *logus* means to study. Meteorology is science deals with the physics, chemistry and dynamics of atmosphere and also their direct and indirect effects upon the earth surface, oceans and thereby life in general.

Climatology: Climatology is scientifically defined as weather conditions averaged over a period of time. Different regions of the world have different characteristic climates which determine the crops of any region. Climatology is derived from Greek word '*klima*' – place/zone or slope of the earth and logs is study. It discovers, describes and interprets the climate on the basis of causes and long term effects on variation in regional and global climate. Climatology is the science which studies average condition of weather or the state behavior of the atmosphere over a place or region for a long period of time.

Agrometeorology: Agrometeorology is the study of weather and use of weather and climate information to enhance crop production. Agrometeorology is defined as the branch of meteorology that deals with relationship between weather and climate on crop, livestock production and soil management.

Use of Agrometeorology in Agriculture

- Regulating the plant growth and development
- Frost prediction and protection of crop plants
- Warning against the forest fires and cyclonic storm/dust storm
- Soil and water conservation
- Planning irrigation
- Controlling insect pests and plant diseases
- Controlling the pollution air and water
- Minimize the losses caused by heavy rain, floods and hail storm
- Regulate the animal growth, byproduct and yield (milk, eggs and meat)

Atmosphere: is defined as the mixture of gases that surrounds the earth or any other star, planet *etc*. About 71 per cent of the earth is made up of water (Hydrosphere) and the remaining is the land portion (Lithosphere). A thick mantle of gases covers the earth and this gaseous portion surrounding the earth is called as atmosphere. The envelope of colourless, tasteless and odourless gasses surrounding the earth is called atmosphere. It is derived from Greek words ***atoms***–water vapour and ***spharia*** – sphere that is sphere of water vapour.

Composition of the Atmosphere

Eight to ten kilometres of air above the earth is essentially a mixture of nitrogen, oxygen, carbon di-oxide, water vapour, dust and several rare gases- Argon, neon, helium and methane. The approximate percentage of these gases except carbon di-oxide and water vapour are practically same everywhere. The percentage of water vapour is quite variable from 0.2 per cent to 2.5 per cent. Nitrogen is four times as that of oxygen.

Constituent	% by Volume
Nitrogen (N_2)	78.08
Oxygen (O_2)	20.95
Argon (Ar)	0.93
Carbon dioxide (CO_2)	0.037

Constituent	Parts per million (ppm) by Volume
Neon (Ne)	18.18
Helium (He)	5.24
Methane (CH_4)	1.70
Krypton (Kr)	1.14
Hydrogen (H_2)	0.60
Nitrous Oxide (N_2O)	0.30
Ozone (O_3)	0.07
Chloro fluoro Carbons (CFCs)	0.0002

Weather and Climate

Weather: State or condition of the atmosphere at a given place and at a given instant of time or State of atmosphere at a particular time in small area as defined by the various meteorological elements - as per World Meteorological Organisation (WMO).

Climate: Generalized weather or summation of weather conditions over a large area during comparatively longer period-as per World Meteorological Organisation (WMO).

Agricultural climate: Agriculture climate is the systematic study of climate of a place or a region in relation to agriculture. It has both temporal and spacial dimensions. Temporal means the changes in weather that occur on day to day or on month to month basis over the year. The temporal dimension of the climate also means the climate is dynamic and it changes from day to day and as such gives rise to seasons. The spacial dimension means the climate conditions vary from place to place.

Important climatic factors for many vital functions of the plant are solar radiation, Precipitation, Temperature, Atmospheric humidity, Wind and Atmospheric gases

Solar radiation: Is defined as the total frequency of electromagnetic spectrum radiation produced by the sun. This spectrum covers visible light and near-visible radiation, such as X-rays, ultraviolet radiation, infrared radiation and radio waves. The visible light and heat of the sun makes life possible and is called daylight or sunshine. The earth's atmosphere deflects or filters the majority of the sun's harmful radiation.

Solar radiation is the source of energy for all physical processes taking place in the atmosphere. This energy also drives the process of photosynthesis, evaporation, heating of soil & air to regulate temperature besides germination, leaf expansion, stem growth, flowering, fruiting *etc.,* More than 70% of solar radiation absorbed by the plant is converted into heat which is essential for transpiration. About 28% of solar energy is utilized in photosynthesis.

A constant amount of radiation (1.94 cal cm^{-2} min^{-1}) is emitted continuously from the sun. The amount of solar energy received per unit area of surface held at right angle to the sun rays at the top of the atmosphere is called **solar constant**.

Solar radiation is the basis for all life on earth. Autotrophs are organisms that produce their own food from the sun (mainly plants) by utilizing solar energy along with carbon dioxide and water to produce simple sugars through the process called photosynthesis. Heterotrophs are organisms that eat other organisms (like animals and fungi) and depend on autotrophs for their food. Heterotrophs can not exist without autotrophs and autotrophs cannot exist without the sun, so life in the universe depends on electromagnetic radiation.

Electromagnetic energy: Electromagnetic energy is the term used to describe all kinds of energies released into space by Sun. Sunlight is a portion of the electromagnetic radiation given off by the Sun in particular infrared, visible and ultraviolet light. On Earth, sunlight is filtered through the Earth's atmosphere and is obvious as daylight when the Sun is above the horizon. When the direct solar radiation is not blocked by clouds, it is experienced as sunshine, a combination of bright light and radiant heat. When it is blocked by the clouds or reflects off other objects, it is experienced as diffused light. Sun takes about 8.3 minutes to reach the earth surface from surface of sun.

Light: Light is the visible portion of the solar spectrum with wavelength range from 0.39–0.76 μm. Light is one of the important climatic factors for many vital functions of the plant. It is essential for the synthesis of the most important pigment that is chlorophyll. Chlorophyll absorbs the radiant energy and converts it into potential energy of carbohydrate. The carbohydrate thus formed is the connecting link between solar energy and living world. In addition, it regulates the important physiological functions.

Light affect plants growth through its intensity, duration and quality

Light intensity: The total quantity of light received in unit time can be considered as light intensity, which is expressed in Foot Candles/Lux. The optimum light intensity for most of the crops is around 1700 Lux. There are certain plants which grow luxuriously with high light intensity; such plants are called **Heliophytes** (Sun-loving) *Example*: Sunflower. The plants growing under low light intensity/shade condition are called **Sciophytes** (Shade loving) *Example*: Coffee.

The light intensity at which the photosynthates produced through photosynthesis is completely made use for its respiration is called **Light saturation point.** It varies from sciophytes (27 Lux) to heliophytes (4200 Lux). Higher light intensity may disintegrate chlorophyll.

Quality of light: The process of photosynthesis in green plants depends on quality of light. Radiation up to 250 nm in UV spectrum is harmful to most of the plants. Though, radiation from 300 nm is photosynthetically active, visible spectrum (400–750 nm) is more effective. Visible spectrum composed with seven colour range VIBGYOR.

Violet	:	400–435 nm
Indigo	:	435–460 nm
Blue	:	460–490 nm
Green	:	490–574 nm
Yellow	:	574–595 nm
Orange	:	595–626 nm
Red	:	626–750 nm

Within the visible range, the principle wavelength absorbed in photosynthesis are Violet to Blue (400–490 nm) and Orange to Red (595–750 nm). Red light is found most favourable. Radiation above 750 nm (Infra-Red radiation) is harmful to photosynthesis.

Duration of light: Light intensity and duration determine the amount of light that a plant receives. The length of the day is more important than intensity. The response (Flowering) of the plants to the relative length of day and night is known as **photoperiodism**. Different species and in some cases different cultivars of the same species responds differently to the length of day and night. The photoperiod required to induce flowering is called **critical day length**.

Based on the photoperiodic response, the plants are grouped as

Long day plants: Crops that reproduce (Flower) normally when the light period is longer than 12 hours. *Example*: *Rabi* Crops-Wheat, Barley, Sugar beet, Potato *etc.*

Short day plants: Crops that reproduce (Flower) normally when the light period is shorter than critical minimum (<12 hours). *Example*: Tobacco, soybean, millets, Rice *etc.*

Day neutral plants: Plants are non-responsive to relative length of day and night. *Example*: Tomato, Sunflower, Buck wheat, Maize *etc.*

Direction of light: The direction of sunlight has a greater effect on the orientation of roots and leaves. In temperate regions, the Southern slopes plants produce better growth than the Northern slopes due to higher contribution of sunlight in the Southern side. The change of position or orientation of organs of plants caused by light is usually called as phototropism that is leaves are oriented at right angles to incidence of light to receive maximum light radiation.

Solar radiation interception and crops production: Dry matter production of crop canopy is directly related to the amount of photosynthetically active radiation (PAR) intercepted by the canopy. The biomass produced (W) can be considered as the product of three factors in the expression.

$$W = \sum Q \times Es \times Ei \times Ec \times dt$$

Q = Daily incident radiation (MJ m^{-2})

Ei = Fraction of incident radiation absorbed by the crop (MJ m^{-2})

Ec = Dry matter/radiation quotient *i.e.,* the efficiency of conversion of radiation into dry matter or radiation utilization efficiency (RUE) in g MJ^{-1}.

Es = Daily quantity of incident PAR on the top of the canopy (MJ m^{-2})

W = Dry matter production (gm^{-2})

dt = Time in days.

Light use efficiency: The conversion efficiency (e) or quantity of dry matter produced per unit light intercepted is known as light use efficiency (LUE).

$$LUE = \frac{\text{Amount of dry matter produced } (gm^{-2})}{\text{Amount of cumulative light absorbed } (MJ\ m^{-2})}$$

Efficiency of solar radiation utilization: Efficiency of solar radiation utilization can be calculated from dry matter production and energy used for it or by comparing calorific value of organic matter produced per unit cultivated land area with the incident solar radiation on the same area during that period. In general less than 1 per cent of solar radiation is utilized by the crops. Efficiency for maize is 1.5 per cent, sugarcane 1.43 per cent, wheat 0.52 per cent, potato 0.5 per cent and sugar beet 0.92 per cent.

Productivity in tropics verses temperate regions: It is often claimed that production potential of cops is more in temperate than tropics due to less biotic and abiotic limitations. In the tropics, high temperature throughout the year and low radiation due to cloudy sky as compared to those in temperate limit crop production. Average radiation in temperate during summer is 1.5 times more than in tropics and the crop growth in temperate is about 1.5 times that in tropics. Rice varieties yield 1.5 to 2 t ha^{-1} as against 4 to 5 t ha^{-1} in temperate. In tropics crops can cultivate throughout the year and the total productivity per unit area is much more in the tropics than that in the temperate.

Precipitation: Precipitation is any product of the condensation of atmospheric water vapour that falls under gravity. The main forms of precipitation include drizzle, rain, sleet, snow, graupel and hail. Precipitation occurs when a portion of the atmosphere becomes saturated with water vapour, so that the water condenses and "precipitates".

Rain: It is precipitation of water in liquid form and drop size is more than 0.5 mm in diameter. The droplets are less close to one another. Rainfall is the most important factors influences vegetation of a place. Most of the plant water demand is met out with rainfall. Rainfall both in total amount and seasonal distribution greatly affect choice of crops and their growth *e.g.*, In heavy and evenly distributed rainfall areas, crops like rice in plains, coffee, tea, rubber in hilly areas can be taken. In low and uneven distributed areas, drought resistance crops like bajra, sorghum, millets can be taken.

Development of rainfall: Rain is liquid water in the form of droplets that have condensed from atmospheric water vapor and then become heavy enough to fall under gravity. Rain is a major component of the water cycle and is responsible for depositing most of the fresh water on the Earth. It provides suitable conditions for many types of ecosystems, as well as water for hydroelectric power plants and crop irrigation. The major cause of rain production is moisture moving along three-dimensional zones of temperature and moisture contrasts known as weather fronts. If enough moisture and upward motion is present, precipitation falls from convective clouds (those with strong upward vertical motion) such as cumulonimbus (thunder clouds) which can organize into narrow rainbands. In mountainous areas, heavy precipitation is possible where upslope flow is maximized within windward sides of the terrain at elevation which forces moist air to condense and fall out as rainfall along the sides of mountains. On the leeward side of mountains, desert climates can exist due to the dry air caused by downslope flow which causes heating and drying of the air mass. The air raises under three processes and based on these processes, there are three different types of rainfall *viz.*, convectional, orographic and cyclonic.

Convectional precipitation: Solar radiation is the main *Source* or heat to produce convection currents in the air. It happened mostly during summer. When the solar radiation is high, thermal low is created at the earth's surface due to high surface temperature. The thermal low results in ascending of air, follows adiabatic cooling and condensation takes place. With enough continuous flow of atmospheric moisture, heavy rainfall expected. The pre-monsoon showers are due to convectional precipitation. These types of precipitation are of a very short duration and consist of heavy showers. The cloud involved in this type of precipitation is cumuliform. They are mostly cumulonimbus.

Orographic precipitation: Mountain or high lands acting as barriers to the flow of air, it forces air to rise and cool adiabatically resulting in formation of clouds. Precipitation occurs due to the above process during the monsoon. The north eastern States, east Rajasthan and Kerala as found along the windward side of mountain ranges, benefiting from orographic rains. Example: southwest monsoon

Cyclonic precipitation: Occurs when deep and extensive air masses are made to converge and move upward so that their adiabatic cooling results. It takes place when cyclone is formed over the sea. In India, the east coast is vulnerable to frequent cyclones while the west coast relatively free. The reason for the above is that the see surface temperature is more and conducive for development of cyclones over the Bay of Bengal when compared to the Arabian Sea. The occurrence of cyclones is common from April to May and from October to November. They give copious rains. At the same time, the cyclones destroy the crops as well as physical infrastructure due to heavy rains and strong winds, causing the loss in millions of Rupees.

Monsoons of India

Monsoon: Monsoon is traditionally defined as a seasonal reversing wind accompanied by corresponding changes in precipitation, but is now used to describe seasonal changes in atmospheric circulation and precipitation associated with the asymmetric heating of land and sea. Usually, the term monsoon is used to refer to the rainy phase of a seasonally changing pattern, although technically there is also a dry phase.

Monsoon is a term derived from the Arabic word *Mausim* meaning season. It was first used by Arabic navigators to describe the seasonal winds of the Arabian Sea. These winds blow from the north-east for one half of the year and from the south-west for the other half.

Indian subcontinent receives its annual rainfall by the peculiar phenomenon known as monsoon. It consists of series of cyclones that arise in Indian Ocean. These travel in northeast direction and enter the Peninsular India along its west coast. The most important of these cyclones usually occur from June to September resulting in southwest monsoon. This is followed by a second rainy season from October to December. A third and fourth rainy seasons occur from January to February and from March to May, respectively. Of the four rainy seasons, southwest monsoon is the most important as it contribute 80 – 95% of the total rainfall of the country.

Significant feature of Indian Monsoon: Climate of India is controlled by the monsoon (season) circulation, literally, the word monsoon can be applied to any wind system in which there is reversal of prevailing wind direction between seasons (summer and winter). Indian subcontinent comes under the influence of southwest (SW) monsoon during June to September and Northeast monsoon (NE) during October to December.

South West Monsoon: Beginning of the year, temperature of the Indian Peninsular rapidly rises under the increasing heat of the sun. A low pressure is established in the interior parts of the Peninsular by the month of March. Westerly winds prevail on the west Kerala and south winds on the Orissa and Bengal. During April and May the region of high temperature is shifted to north *viz.,* Upper Sindh, lower Punjab and Western Rajasthan. This area becomes the low pressure area to which monsoon winds are directed.

The western branch of South West monsoon touches North Karnataka, Southern Maharashtra and then it make its way to Gujarat. When the South West Monsoon is fully operating on the Western India, another branch of the same is acting in the Bay of Bengal. It carries rains to Burma, Northern portions of the east coast of India, Bengal, Assam and the whole of North India in general.

North east Monsoon: During September end, the South West Monsoon penetrates to North Western India but stays on for a full month in Bengal. On account of the increase in pressure in Northern India, there is a shift of the pressure to the South East and North Easterly winds begin to flow on the eastern coast. These changes bring on heavy and continuous rainfall to the Southern and South Eastern India.

Winter Rainfall: It is restricted more to Northern India and is received in the form of snow on the hills and as rains in the plains of Punjab, Rajasthan and central India. Western disturbance is a dominant factor for rainfall during these months in north western India.

Summer Rainfall: It is received from March to May as local storms. It is mostly received in the South East of Peninsular and in Bengal. Western India does not generally receive these rains.

Isohyets: An isohyet is also known as isohyetal line, and it is a line on a map which connects points that have the same amounts of precipitation in a given period or for a particular storm. Isohyets are then drawn to create an isohyetal map.

Atmospheric temperature: Temperature is a relative term implying a degree of molecular activity or heat of a substance. Heat is the energy arising from random motion of all the molecules in a body. The primary *Source* of heat for the earth and its atmosphere is the sun. The heat energy is transmitted by three mechanisms *viz.,* conduction, convection and radiation.

Isotherms: The lines joining the places of equal temperature at a given date or time on a geographic map.

Factors Affecting Air Temperature

1. Latitude
2. Altitude
3. Distribution of land and water
4. Ocean currents
5. Prevailing winds
6. Cloudiness
7. Mountain barriers

1. **Latitude:** The time of occurrence of maximum monthly mean temperature and minimum monthly mean temperature also depends on latitude of a place. *Eg.* The coldest month is January in northern regions of India while December in the south. Similarly, the warmest month is May in the south while June in the north across the country.
2. **Altitude:** The surface air temperature decreases with increasing altitude from the mean sea level as the density of air decreases. Since the density

of air is less at higher altitudes, the absorbing capacity of air is relatively less with reference to earth's long wave radiation.

3. **Distribution of land and water:** Land and water surfaces react differently to the insolation. Because of the great contrasts between land and water surfaces their capacity for heating the atmosphere varies. Variations in air temperature are much greater over the land than over the water. The differential heating process between land and sea surfaces are due to their properties. It is one of the reasons for Indian monsoon.
4. **Ocean currents:** The energy received over the ocean surface carried away by the ocean currents from the warm areas to cool areas. This results in temperature contrast between the equator and poles. The occurrence of El-Nino is due to change in sea surface temperature between two oceanic regions over the globe.
5. **Prevailing winds:** Winds can moderate the surface temperature of the continents and oceans. In the absence of winds, we feel warm in hot climates. At the same time, the weather is pleasant if wind blows.
6. **Cloudiness:** The amount of cloudiness affects the temperature of the earth's surface and the atmosphere. A thick cloud reduces the amount of insolation received at a particular place and thus the day time temperature is low. At the same time, the lower layers in the atmosphere absorb earth's radiation. This results in increasing atmospheric temperature during night. That is why, cloudy nights are warmer. This is common in the humid tropical climates.
7. **Mountain barriers:** Air at the top of the mountain makes little contact with the ground and is therefore cold while in the valley at the foothills makes a great deal of contact and is therefore warm. That is, the lower region of the earth's atmosphere is relatively warmer when compared to hillocks.

Influence of Temperature in Crop Production

1. Temperature influences distribution of crop plants and vegetation.
2. The surface air temperature is one of the important variables, which influences all stages of crop during its growth development and reproductive phase.
3. Air temperature affects leaf production, expansion and flowering.
4. The diffusion rates of gases and liquid changes with temperature.
5. Solubility of different substances is dependent on temperature.
6. Biochemical reactions in crops (double or more with each 10°C rise) are influenced by air temperature.

7. Equilibrium of various systems and compounds is a function of temperature.
8. Temperature affects the stability of enzymatic systems in the plants.
9. Most of the higher plants grow between 0°C-60°C and crop plants are further restricted from 10–40°C, however, maximum dry matter is produced between 20 and 30°C.
10. At high temperature and high humidity, most of the crop plants are affected by pests and diseases.
11. High night temperature increases respiration and metabolism.
12. A short duration crop becomes medium duration or long duration crop depending upon its environmental temperature under which it is grown.
13. Most of the crops have upper and lower limits of temperature below or above which, they may not come up and an optimum temperature when the crop growth is maximum. These are known as cardinal temperatures and different crops have different temperatures.

Heat Units Concept

The relationship between the amount of heat and plant was first quantified by Beaumur in 1735 by summing up the mean daily temperature above base temperature during each plant development phase. This concept assumes that there is direct relationship between plant growth and temperature. According to this concept, higher the daily temperature is shorter the duration. Heat unit concept is also called heat sums, day degrees, growing degree days (GDD) or degree days is relative warmth of growing season of a given duration.

Heat unit is the departure from the mean daily temperature above the minimum threshold temperature. Heat unit or GDD is calculated as indicated below;

$$GDD = \sum_{r=1}^{n} \left\{ \frac{T_{\max} + T_{\min}}{2} \right\} - T_b$$

Where,

T_{max} : Maximum air temperature of the day

T_{min} : Minimum air temperature of the day

T_b : Base temperature of the crop

The base temperature is also called threshold temperature is the temperature below which no growth takes place. It varies with the crops, usually higher values for tropical crops and lower values for temperate crops.

Crops	Threshold temperature (°C)
Wheat and barley	4.4
Potato, oats, sugar beet	5.0
Sorghum, maize, groundnut	8–10
Rice	10–12
Tobacco	13–14

Advantages/Importance of growing degree day concept

1. In guiding the agricultural operations and planning land use
2. To forecast crop harvest dates, yield and quality
3. In forecasting labour required for agricultural operations
4. Introduction of new crops and new varieties in new areas
5. In predicting the likelihood of successful growth of a crop in an area

Atmospheric Humidity

Humidity refers to the water vapour suspended in the atmosphere. Water vapour is a colourless, odourless, tasteless and consists of very minute water droplets suspended in air. This is the resultant of the evaporation from soil and water bodies with the solar energy in the form of temperature. When the atmosphere holds maximum amount of water vapour at particular temperature is said to be saturated. Increase in temperature will make the air unsaturated. The evapo-transpiration of water is directly proportional to the difference between the amount of water actually present and the saturation. The difference is termed as **Vapour pressure deficit (VPD).**

The atmospheric humidity is expressed as absolute, specific and relative humidity.

Absolute humidity is the actual quantity of water vapour by weight present in a given volume of air

$$\text{Absolute humidity (gm}^{-3}\text{)} = \frac{\text{Weight of water vapour (g)}}{\text{Volume of air (m}^{3}\text{)}}$$

Specific humidity is the weight of water vapour per unit weight of air (including water vapour)

$$\text{Specific humidity (g kg}^{-1}\text{)} = \frac{\text{Weight of water vapour (g)}}{\text{Weight of air including water vapour (kg)}}$$

Absolute and specific humidity are difficult to measure as they fluctuate greatly with temperature. Also, they have little or no effect on crop growth.

The extent of evaporation and transpiration depends on the amount of moisture held in the atmosphere. Hence, it is essential to find out the saturation capacity and VPD, which is given by relative humidity.

Relative humidity is the ratio between the amount of water vapour present in the air and the amount of vapour required for saturation at particular temperature and pressure and expressed in percentage

$$\text{Relative humidity (\%)} = \frac{\text{Water vapour present in air}}{\text{Water vapour required for saturation}}$$

Relative humidity is greatly influenced by temperature, wind, vegetation and water content in soil.

Effect of relative humidity on crop

1. If RH is 100%, there is no evaporation and transpiration and if it is 0%, then there is enormous evapotranspiration. Both these conditions are not congenial for crop growth. RH of around 60% is suitable for most of field crops.
2. Increased RH (> 80%): Favours growth of many fungi and bacteria, affect gas exchange through stomata in turn crop growth and development. High RH can prolong the survival of crops under soil moisture stress. Certain plants have the ability to absorb atmospheric water vapour through leaves and other aerial parts under situations of high RH.
3. Reduced RH (< 40%): Increase evapo-transpiration and demand more irrigation.
4. Optimum relative humidity of 40–60% is congenial for crop growth and development.

Relativc humidity across the country depends on temperature and moisture contents of the atmosphere, RH in the country exhibits seasonal variation depending on monsoon air mass movements. In the central and western parts of the country, RH values are lowest 20–40% during summer period. Western part of Rajasthan record lowest RH (around 4%). With approaching monsoon by June, the mean RH in almost all the locations in the country ranges between 50–80 per cent. Maximum RH is in the equatorial region due to high evaporation. Decrease towards poles upto 30° N and S due to subsiding air mass. RH increases in poles due to low temperature.

Wind: Air in horizontal movement is known as wind. Air moves constantly from place to place due to the pressure difference and which is caused due to difference in temperature. Winds are named after the direction from it blows. Vertical moving air columns are called currents and are negligible and vertical movement or uplift of air only causes significant weather changes in cloud formation and rain.

Effect of Wind on Crop Plants

1. Transports heat in either sensible or latent heat form from lower to higher altitudes.
2. Wind affects the plant directly by increasing transpiration and the intake of CO_2 and also causes several types of mechanical damage.
3. Wind helps in pollination and dispersal of seeds.
4. Light and gentle winds are helpful for cleaning the agricultural produce.
5. Hot dry winds frequently do much damage to vegetation in the growing crops by promoting excessive water loss.
6. Wind has powerful effect on humidity.
7. Long, continued warm, dry winds injure blossoms by evaporating the secretion of the stigma.
8. Provides moisture which is necessary for precipitation.
9. Wind prevents frost by disrupting atmospheric inversion.
10. Causes soil erosion.

Atmospheric gases: Atmospheric air is composed of 78.09% Nitrogen, 20.98% Oxygen, 0.93% Argon & 0.03% CO_2. CO_2 is most important for photosynthesis and taken by the plant through diffusion in stomata. Oxygen is important in respiration of both plants & animals. Nitrogen is important as plant nutrient. Atmospheric nitrogen is fixed in the soil by lightening, rainfall & microbes. Certain gasses like SO_2, CO, CH_4, HF, CFC released from petrochemicals are toxic.

Isobars: A line connecting points of equal pressure is called an **isobar**. The **isobars** are generated from mean sea level pressure reports and the pressure values are given in millibars. Any point in between these two **isobars** will have a pressure somewhere between 996 mb and 1000 mb.

Plant Weather Modification

Plants have a remarkable ability to cope with diverse environments. However, the whims of man can lead to plants being so misplaced that environmental modification may be a useful enterprise for improving productivity. Unfortunately, the potential for modifying canopy environments on large scales under drought conditions is limited.

There are different ways by which one can modify the plant environment, they are:

1. Artificial rain making
2. Heat transfer

3. Controlling heat load
4. Heat trapping and shading
5. Protection from cold injury
6. Controlling soil moisture

Artificial Rain Making

Cloud seeding: Cloud seeding is one of the tools to mitigate the effects of drought. It is defined as a process in which the precipitation is encouraged by injecting artificial condensation nuclei through aircrafts or suitable mechanism to induce rain from rain bearing cloud. The rain drops are several times heavier than cloud droplets. These mechanisms are different for cold and warm clouds.

Seeding of cold clouds: This can be achieved by two ways

1. Dry ice seeding (solid carbon-dioxide)

- Dry ice remains as it is at –80°C and evaporates, but does not melt. Dry ice is heavy and falls rapidly from top of cloud and has no persistent effects due to cloud seeding.
- Aircrafts are commonly used for cloud seeding with dry ice. Aircraft flies across the top of a cloud and 0.5–1.0 cm dry ice pellets are released in a steady stream.
- While falling through the cloud a sheet of ice crystals is formed.
- From these ice crystals rain occurs.

This method is not economical as 250 kg of dry ice is required for seeding one cloud. To carry the heavy dry ice over the top of clouds special aircrafts are required, which is an expensive process.

2. Silver Iodide seeding

Minute crystals of silver iodide produced in the form of smoke acts as efficient ice-farming nuclei at temperatures below –5°C. When these nuclei are produced from the ground generators, these particles are fine enough to diffuse with air currents. Silver iodide is the most effective nucleating substance because; its atomic arrangement is similar to that of ice. The time for silver iodide smoke released from ground generator to reach the super cooled clouds was offer some hours, during which it would draft a long way and decay under the sun light. The appropriate procedure for seeding cold clouds would be to release silver iodide smoke into super cooled cloud from an aircraft. In seeding cold clouds silver iodide technique is more useful than dry ice techniques, because, less silver iodide is sufficient per cloud.

Seeding of warm clouds

1. **Water drop Technique:** Coalescence process is mainly responsible for growth of rain drops in warm cloud. The basic assumption is that the presence of comparatively large water droplets is necessary to initiate the coalescence process. So, water droplets or large hygroscopic nuclei are introduced in to the cloud. Water drops of 25 mm are sprayed from aircraft at the rate of 30 gallons per cloud.
2. **Common salt technique:** Common salt is a suitable seeding material for seeding warm clouds. It is used either in the form of 10 per cent solution or solid. A mixture of salt and soap avoid practical problems. The spraying is done by power sprayers and air compressors or even from ground generators. The balloon burst technique is also beneficial. In this case gun powder and sodium chloride are arranged to explode near cloud base dispersing salt particles.

Heat Transfer

Transpiration: Transpiration consists of the vaporization of liquid water contained in plant tissues and the vapour removal to the atmosphere. Crops predominately lose their water through stomata. These are small openings on the plant leaf through which gases and water vapour pass. The water, together with some nutrients, is taken up by the roots and transported through the plant. The vaporization occurs within the leaf, namely in the intercellular spaces, and the vapour exchange with the atmosphere is controlled by the stomatal aperture. Nearly all water taken up is lost by transpiration and only a tiny fraction is used within the plant. When water vapour is transpired from leaves into the atmosphere, the plant and surrounding air is cooled down.

Controlling Heat Load

The heat load on the plant can be controlled by

1. Shade of a nurse plant
2. The whitish, silvery, and other metallic finishes of leaves-Desert plants.

Heat Trapping and Shading

Since heat on the surface of the Earth comes from sunlight, "heat" is trapped inside the atmosphere when rays of sunlight make it through our atmosphere, but are not able to make it back out because there are too many greenhouse gases in the atmosphere. Growing of nurse plant can reduce the possible effect of heat load on the crops.

Protection From Cold Injury

To protect the plants from cold injury, apply a layer of 6–8 inches of wood chips or straw over perennial and flower beds. Wet soils prevent freeze injury to roots because moist soil holds more heat than dry soil.

Controlling Soil Moisture

The following practices might be followed to conserve the soil moisture and to modify the plant weather modification

1. Weed control
2. Soil aeration
3. Mulching
4. Conservation methods for crop land
   - Dead furrows or Idle furrows
   - Ridging
   - Graded furrows
   - Broad bed and furrows
   - Tie ridging
   - Mulching
   - Scooping
   - Off season tillage
   - Compartment bunding
   - Zing terraces
   - Contour cultivation
   - Contour ploughing

Remote Sensing and Its Role in Agriculture

The science of acquiring information about an object, without entering in contact with it, by sensing and recording reflected or emitted energy and processing, analysing, and applying that information.

Types of remote sensing

1. Passive remote sensing
2. Active remote sensing

Application of Remote sensing in Agriculture

- **Identification, area estimation and monitoring:** The specific requirement of climate and soil conditions coupled with the specialized management practices make the distribution of plantation crops rather more localized in comparison to other agricultural crops. The identification, estimation of growing stock, analysis of distribution and monitoring at regular intervals are major aspects in plantation crops.

- **Crop nutrient deficiency detection:** The nutrient deficiency in plants affects the colour, moisture content and internal structures of the leaves and as a result their reflecting power changes.
- **Soil mapping:** Advancements in space technology opened application possibilities of remote sensing in soil mapping. Soil properties have also been inferred from optical and microwave data using physically-based and empirical methods. Soil properties that have been measured using remote or proximal sensing approaches include mineralogy, texture, soil iron, soil moisture, soil organic carbon, soil salinity and carbonate content.
- **Crop condition assessment:** The physiological changes that occur in a plant due to stress may change the spectral reflectance characteristics resulting in the detection of stress amenable to remote sensing techniques. Crop monitoring at regular intervals during the crop growth cycle is essential to take appropriate measures and to asses information on probable loss of production.
- **Agricultural drought assessment:** Drought assessment is yet another area wherein remote sensing data has been used at operational level. The district level drought assessment and monitoring using NDVI generated from NOAA - AVHRR data helps in taking timely preventive and corrective measures for combating drought.
- **Reflectance modelling:** Physical reflectance models for crops serve the important purpose of understanding the complex interaction between solar radiation and plant canopies. In order to obtain a reliable yield prediction, growth of crops has to be modelled by means of crop growth models. Crop growth models describe the relation between physiological process in plants and environmental factors such as solar radiation, temperature, water and nutrient availability.
- **Crop yield modelling and production forecasting:** The information on production of crops before the harvest is very vital to the national food policy planning and economy of the country. Reliable crop yield estimate is one of the most important components of crop production forecasting.

Table 9: Satellites used for remote sensing

S. No.	Satellites	Year	Sensors used
1.	IRS-1A/1B	1988, 1991	LISS 1 and LISS-II
2.	IRS-P2	1994	LISS-II
3.	IRS-1C/1D	1995, 1997	LISS-III, WiFS and PAN*
4.	IRS-P3	1996	MOS-A, MOS-B MOS-Cand WiFS**
5.	IRS-P4	1999	OCM and MSMR
6.	IRS-P6	2003	Liss-IV, LISS-111 and AWiFS**
7.	IRS-P5(Cartosat-1)	2005	PAN ($+26^0$) & Aft (-5^0)
8.	Cartosat-2	2007	PAN
9.	Cartosat-3	2019	PAN

Rainfall pattern, rainfall characteristics and rainfall variability analysis

Rainfall pattern in India

India has a monsoon type of climate. Rainfall occurs only for about three to four months and is uneven, unreliable and erratic in nature. Rainy season in India extends from June to mid- September. Sometimes at the time of arrival, the rainfall pours heavily which is commonly known as the burst of the monsoons. The monsoon first arrives at the southern tip of the subcontinent. It then gets separated into two branches; the Arabian Sea branch and the Bay of Bengal branch. The rains from the Arabian Sea strikes Mumbai on approximately June 10, while the Bay of Bengal branch brings rainfall to Assam. Because the mountains in the northeast causes the rain bearing winds to drift to the Indo-Gangetic plains, the region experiences rainfall by late June and early July. The rainfall reaches Himachal Pradesh by mid July. While it rains heavily in north eastern parts of the country, Kerala and on the western slopes of the Western Ghats, moderate rainfall is experienced in the south eastern parts of the country and the Indo Gangetic Plains. Scanty rainfall is experienced in western parts of Rajasthan, Punjab, Haryana and Gujarat.

Rainfall and dry land agriculture in India

Dryland areas are characterized by low and erratic rainfall and no assured irrigation facilities. Dryland agriculture is important for the economy as most of the coarse grain crops, pulses, oilseeds, and raw cotton are grown on these lands. Dryland areas receive rainfall between 500 and 1200 mm. The fate of dry land agriculture depends on behaviour of Southwest monsoon in India, which contributes 80 per cent of total rainfall. Onset of monsoon, its distribution and the quantity of rainfall largely influence dryland crop production. Major deviations from normal performance of rainfall pattern may lead to economic loss or even to distress conditions in India.

Intensity of Rainfall

The amount is of course important to the overall hydrologic cycle and replenishment of the soil water, and the amount is an accumulation or product of the intensity times the duration. For example, the amount may be the same for a high intensity short duration rainfall as it is for a low intensity long duration rainfall. However, the intensity and duration can have a large influence on whether the rainfall infiltrates or becomes surface runoff. Higher rainfall intensity produces larger size raindrops which have more impact energy, and thus higher intensity storms can damage delicate vegetation and bare soil. High intensity storms can literally displace soil particles, causing soil crusting or starting the soil erosion process. High intensity storms may also overwhelm the soils ability to infiltrate the rainfall at the same rate, causing infiltration-excess runoff.

Rainfall Characteristics

There are following rainfall characteristics

1. Onset of effective monsoon
2. Withdrawal of effective monsoon
3. Length of rainy season
4. Dry spells
5. Wet spells

Onset of Effective Monsoon (OEM)

Over India, monsoon onset occurs initially across the south peninsula in early June, when heavy rains lash south peninsula after the cross-equatorial low-level jet (LLJ) is established across the Somali coast into the near-equatorial Arabian Sea. By middle of July, monsoon covers the whole country.

The concept developed by Ashok Raj (1979), for onset of effective monsoon (OEM) was followed. According to this concept the date of commencement of seven days spell satisfying following three criteria is taken as the date of onset of effective monsoon

- The first days rain in the seven day spell should be more than the average daily evaporation (e) mm of the place.
- The total rain during the seven days spell should be at least (5e + 10 mm).
- At least four out of these seven days should be rainy day having rainfall more than or equal to 2.5 mm.

Withdrawal of Effective Monsoon (WEM)

During the monsoon period there are some dry spells out of which some may be critical dry spells (CDS). Generally the monsoon period is from June to

September. At the end of the monsoon season the dry spell of relative longer period is observed. By analyzing the daily rainfall data, the dry spell at the end of the monsoon season is determined also there is drastic reduction in the weekly rainfall at the end of monsoon season. Based on this principle, the dates of withdrawal of monsoon were computed as per the produce adopted in case of OEM.

Length of Rainy Season

Length of rainy season is the difference in days between the onset of effective monsoon and withdrawal of monsoon. It helps in selecting different crops, varieties and cropping pattern for particular region.

Dry Spells

From the agronomic point of view a dry spell is not a drought of climatic magnitude but a period of a few days or weeks without rainfall. A period of atleast 15–20 days is considered as dry spell period without rainfall which creates moisture stress and yield may be reduced.

Critical dry spell

A critical dry spell is defined as the duration between the end of a wet spell and the start of another wet spell; during which a 50 per cent depletion of available soil moisture occurs in the top 50 cm soil layer. It is determined by the following relation:

$$\text{CDS} = \frac{\text{AMD}}{\text{ET}}$$

CDS = Critical dry spell

AMD = 50 per cent of the available soil moisture in the top 50 cm soil layer, expressed in terms of depth, (mm) and

ET = Average maximum daily evapotranspiration of a crop (mm/day).

Wet Spell

A period of a number of consecutive days on each of which precipitation exceeding a specific minimum amount has occurred. Wet spell (week) may be called as wet week if rainfall received during this week was more than 20 mm (Pandrinath, 1991). 6.3 mm per day is called as Agricultural rainy day and 5 mm per day is crop rainy day.

Calculation of Effective Rainfall

Total rainfall can be measured directly with rain gauges. Several types of recording and non-recording gauges are available. The instruments and their use have been described by WMO (1970). Irrigation applications can be measured with various types of notches, weirs, Parshall flumes, cut-throat flumes and with direct recording water meters. The methods have been compiled and described by Dastane (1972), With efficient farm water management, surface run-off should be kept to a minimum. Only excess water should be removed by deep percolation and sub-surface drainage or by surface drainage. For a given field, run-off can be assessed with measuring flumes and water stage recorders or can be computed with formulae established for different conditions, as given in the Handbook on Hydrology by Chow (1964), and USDA Field Manual for Research in Agricultural Hydrology (1968).

There are several methods of assessing effective rainfall. Each method has certain merits and limitations

Empirical methods of determining effective rainfall

1. Soil Moisture Changes

Water in the root zone may be measured by sampling and oven-drying the soil before and after every shower of rain. The increase in soil moisture, plus evapotranspiration loss (ETa) from the time the rain starts until the soil is sampled, is the amount of effective rainfall. After heavy rainfall evapotranspiration can be assumed to be at the potential rate during the short period from cessation of rainfall until the sampling time. This can be taken as 0.4–0.8 times the evaporation value of the Class A Pan as is given in FAO Irrigation and Drainage paper No. 24 (1974), or

$$ER = M2 - M1 + kp\ Eo$$

ER = Effective rainfall

Eo = U.S. Class A Open Pan evaporation value

M1 and M2 = moisture status in the effective root zone before and after rain, respectively

kp = Pan coefficient

The method takes into account the soil and the crop characteristics. The determination is simple and accurate but it may involve errors due to soil variation; the sampling errors may range from 5–40 per cent. The method is also laborious and time consuming. The use of neutron probes reduces the drudgery of periodic soil sampling, but these are costly methods for routine purposes and also subject to sampling errors.

2. Daily Soil Moisture Balance Method

A daily soil water balance is rather like a bank account. Rainfall and irrigation are on the credit side, while soil moisture depletion is on the debit side. Precise data on the maximum water holding capacity (field capacity) is necessary for this method. Any amount in excess of this capacity is a surplus and will be a deep percolation loss or run-off. When the balance reaches nil, no more withdrawal is possible and hence further depletion is treated as water deficiency. Rainfall and irrigation are directly measured while the evapotranspiration is computed from any of several available formulae.

In irrigated agriculture, the soil water content is never allowed to fall below a certain value where water becomes a limiting factor in crop production. When water is depleted to the lower limit of readily available moisture, irrigation is applied. Hence, computations may be based on potential evapotranspiration (ETp). In rainfed or partially irrigated areas, where the soil moisture is depleted below the lower limit of readily available moisture, the computations are to be based on actual evapotranspiration. ETa can be estimated by using the Thornthwaite and Mather method (1955), that of Baier and Robertson (1966), or the relationships between ETa and ETp under decreasing soil water content as given by Tanner (1967).

3. Integrating Gauge

The method is shown in Fig 1. It consists of a rainfall receiver (R) which is connected to a water reservoir (WR), and which in turn is connected to an evaporating surface (E) representing a crop. The reservoir is provided with an overflow outlet at the top of the side wall (D). The capacity of the reservoir is adjusted to the maximum water holding capacity of the soil in question. The rain water above this maximum capacity flows out and is measured as ineffective rainfall. The evaporating surface loses moisture continuously, creating a fall in the water level in the reservoir, which is graduated so that the moisture balance can be read directly at any time.

The device is simple, practical and useful and can easily be set up in the field. The evaporating surface represents the crop and therefore its size and the porosity of the material are important. The method is described by Stanhill (1958).

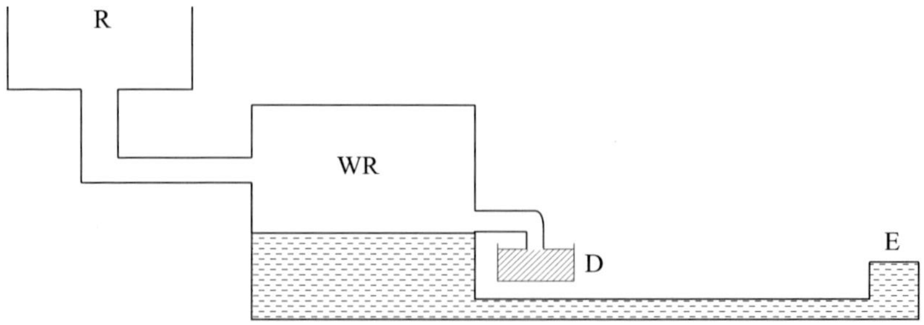

Fig. 1. Integrating Gauge for Measuring Effective Rainfall

4. The Ramdas Method

Ramdas (1960) suggested a direct field method using a small portable device containing soil of the field, so eliminating the necessity of sampling.

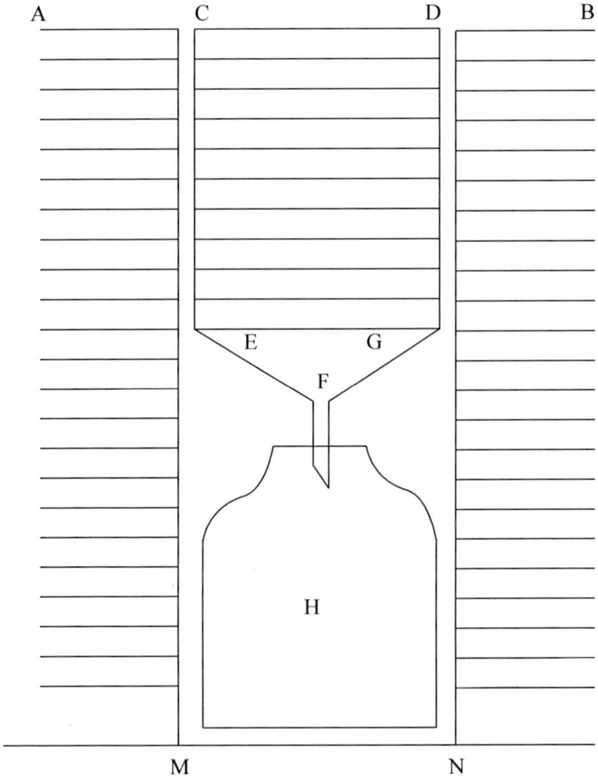

Fig. 2. Ramdas Apparatus for Measuring Effective Rainfall

The apparatus, as shown in Fig. 2, consists of a cylinder (CD) of about 30 cm in diameter, with a perforated base (EG) and a funnel (F) leading into a receiver bottle (H). All these parts are enclosed in an outer cylinder (MN). The cylinder (CD) is filled with a representative soil with the same density as that of the field. The height is equal to the depth of the effective root zone of the crop. The apparatus is installed in the field crop where the effective rainfall is to be measured. The crop in the container is irrigated along with the field crop. Excess rain or irrigation water drains in the receiver bottle H and is measured from time to time. The total rainfall minus the ineffective rainfall gives the value of effective rainfall. It is assumed that there is no surface run-off. Cylinders of different lengths are used consistent with the rooting depth of the different crops. With a suitable number of replications, the method is very useful. It is simple and practical, and furnishes direct readings.

5. Lysimeters

Lysimetry is a method which provides complete information on all the components of water balance. Lysimeters can be used not only for measuring evapotranspiration but also for checking empirical formulae for computing ET. The method is similar to the Ramdas method, but is more elaborate, refined and gives a higher accuracy.

A lysimeter is a large container with soil in which crops are grown; water losses and gains can be measured. The container is fitted with suitable inlets for irrigation and outlets for drainage. The lysimeters are buried in the field and are surrounded by the same crop as is grown inside. The size of lysimeter varies from small oil drums (Gilbert and van Bavel, 1954) to large size and deep lysimeters (Harrold and Dreilbelbis, 1958, 1967 Pruitt and Angus, 1960, McIlroy and Angus, 1963). They can be either the non-weighing or weighing type.

In non-weighing lysimeters, changes in water balance are measured volumetrically weekly or biweekly. No accurate daily estimates can be obtained. A simple design is shown in Fig. 3. Irrigation water is applied to the lysimeter, A layer of pebbles is placed at the bottom to facilitate easy drainage. Excess water is collected from below at a suitable distance. A number of crops can be grown in a concentric pattern around a central drainage chamber. A simple lysimeter can be built at low cost from a petrol drum. A tube with a small diameter is placed through the soil to the layer of pebbles. Excess water is removed at frequent intervals by using a thin metal tube open at the bottom which is connected to a receiver bottle in which suction can be applied using a reversed handpump. This is shown in Fig. 4.

Weighing lysimeters can provide precise information on soil moisture changes for daily or even hourly periods. The lysimeter is placed inside another tank which is in contact with the surrounding soil. The inside container is free for weighing by scales. Also, the lysimeter tank can be floated in water; a suitable heavy liquid ($ZnCl_2$) is used whereby the change in liquid displacement is a measure for the water gain or loss to or from the lysimeter tank. Apart from the high cost, the major problems with lysimeters are the restricted root growth, the disturbed soil structure in the lysimeter causing changes in water movement and possibly the tank temperature regime, resulting in condensation of water on the walls of the container. Harrold and Dreilbelbis (1967) estimated that errors due to dew formation were in the order of 250 mm per annum. Other limitations include the 'bouquet effect' whereby the canopy of the plants grown in the lysimeter is above and extends over the surrounding crop, resulting in a higher evapotranspiration rate. In spite of these limitations, it is the best technique for precise studies on evapotranspiration.

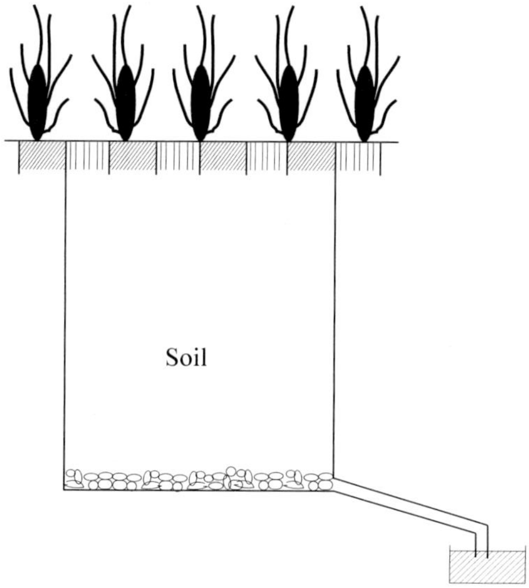

Fig. 3. Drainage Lysimeter

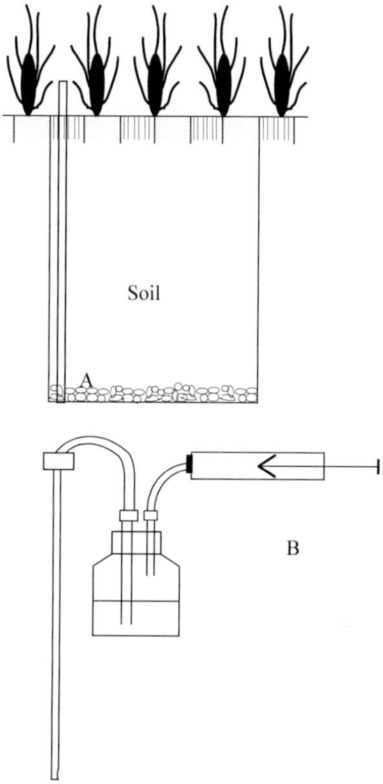

Fig. 4. Suction type Drainage lysimeter

6. Determining effective rainfall from formulae

A number of empirically determined formulae can be used. They have been developed under a given set of conditions which may be very different from those under which they are to be applied. Their use elsewhere therefore remains doubtful.

(i) Renfro Equation

Renfro, as quoted by Chow (1964), suggested the following equation for estimating effective rainfall:

ER = E Rg + A

ER = Effective rainfall

Rg = Growing season rainfall

A = Average irrigation application

E = Ratio of consumptive use of water (CU) to rainfall during the growing season (Table 10).

Table 10: Ratio E for use in estimating effective rainfall in renfro equation

CU/Rg	E	CU/Rg	E
0	0	2.4	0.72
0.2	0.10	2.6	0.75
0.4	0.19	2.8	0.77
0.6	0.27	3.0	0.80
0.8	0.35	3.5	0.84
1.0	0.41	4.0	0.88
1.2	0.47	4.5	0.91
1.4	0.52	5.0	0.93
1.6	0.57	6.0	0.96
1.8	0.61	7.0	0.98
2.0	0.65	9.0	0.99
2.2	0.69	/	/

Source: FAO

The E value implies degree of rain likely to be utilized in meeting consumptive water needs. The greater the E value, the higher the value of effective rainfall. For example, if rainfall during the four month growing season is 400 mm, consumptive use of water is 700 mm, and average irrigation application is 100 mm, then the effective rainfall is equal to $0.60 \times 400 + 100 - 340$ mm. The method is empirical and may not suit many situations.

(ii) Empirical Relationships

Several methods of estimating effective rainfall for irrigation schedules are in vogue in different countries. They are based on long experience and have been found to work quite satisfactorily in the specific conditions under which they were developed.

(iii) Crops other than rice India

For a given area, effective rainfall is taken to be equal to 70 per cent of the average seasonal rainfall. In another method, effective rainfall is taken as the mean value of rain, with the excess ever 3 inch in one day and 5 inch in 10 days omitted. Effective rainfall has also been taken to be equal to the lowest monsoon rainfall occurring in three out of four years. The Damodar Valley Corporation divided the *'Kharif* season' (June to November) into ten-day periods. During these periods, rainfall which exceeded the water needs of the predominant crop was considered as ineffective and the rest as effective. Khushlani (1956) suggested that rainfall during the life cycle of the crop in a bad year (one with low rainfall) should be considered as the effective rainfall. In determining the water requirement of sugar cane under Bombay Deccan conditions, Rege *et al*., (1943) considered that rainfall received only after five days from the irrigation date was effective. The rainfall received within five day after irrigation was treated as ineffective.

Sastry (1956) suggested the following equation for estimating rainfall usable by crops under Andhra Pradesh (central part of South India) conditions,

$$Y = \overline{X} - Cd$$

Y = Usable rainfall, per days for a given period

$\overline{X}$ = Mean daily rainfall

C = A constant, which determines the confidence limit of the minimum statistical average

d = Standard deviation of the daily rainfall

Estimates of usable rainfall in various periods of the monsoon season can be arrived at by selecting suitable confidence limits and statistical methods for calculating the standard deviation.

Rainfall variability analysis and interpretation

The important characteristics of rainfall influencing production of rainfed farming are number of rainy days, drought, normal and excessive rainfall for week, month and year. Meteorologically a day is considered dry when it receives rainfall less than 2.5 mm, while agriculturally a dry day receives rainfall less than 6.3 mm are of no use for plant growth, as they will not wet the soil enough to supply moisture around roots (Chowdhary, 1979). Karate and Sena (2004) studied the application of rainfall analysis for planning soil and water conservation structures in semi-arid Gujarat. Researcher have made meteorological drought, agriculture drought based on rainfall data for different regions in India.

Rainfall variability analysis is important for planning and execution of agricultural operations, for predicting drought/flood and to give fore warning to tide over the situation and for reducing the risks in dry farming. The average annual rainfall helps to decide the crops and cropping pattern, Seasonal rainfall analysis helps to decide the cropping period, Monthly rainfall analysis helps to decide length of growing season and weekly rainfall analysis helps to find out dry and wet spells within the growing season, to plan agricultural operations, to identify mid season risks and to initiate corrective measures.

I. Rainfall Periodicities and A Long Term Trend

1. A graph can be plotted with the year on X-axis and the annual rainfall on the y-axis. The rainfall undergoes a periodic variation with 11-year cycle. This is due to the influence of the sunspot cycle or solar cycle. Percentage deviation from the normal on this graph helps to classify a year as excess (>+ 20%), normal (–19 to 19) or deficit (<–20% of average rainfall) year (Fig. 5).
2. Trend is a long time gradual increase or decrease in a weather parameter. To study the trend the short term fluctuation of annual rainfall due to the influence of solar cycle has to be filtered. To achieve this, 11 year moving average method is used; Graph helps to understand the trend.

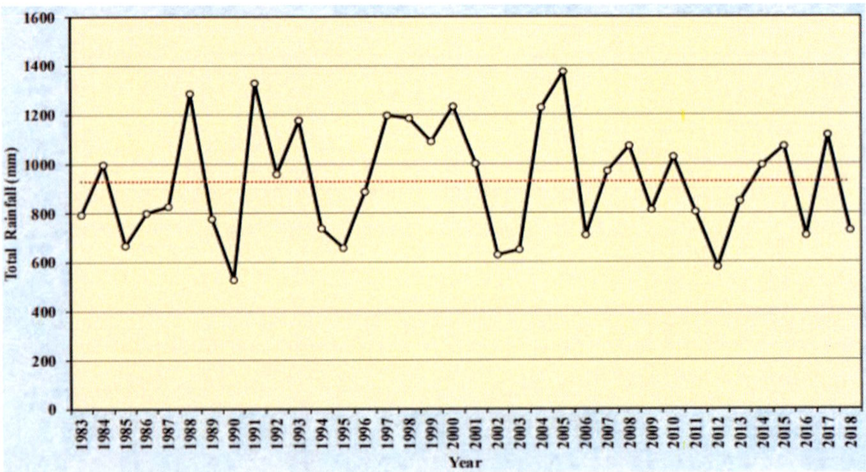

Fig.5. Annual total rainfall of GKVK, Bengaluru from 1983–2018

3. Intensity of rainfall mainly influence erosion of soil. Study of rainfall intensity helps in identifying probable period of floods, filling of irrigation tanks *etc.* If the intensity of rains exceeds the rate of infiltration of soil, run off occurs. Intensity of rainfall in mm/day in the last 10 years is considered to estimate run off losses.

4. Number of rainy days (days with > 2.5 mm rain) indicates distribution of rainfall in the growing season and also length of growing season. A rainfall < 0.5 mm is recorded as traces and not considered for accounting.
5. Periodicity of rainfall : Certain events of rainfall are repeated at periodical intervals such events are onset of monsoon, recurrence of droughts, probability of receiving certain amount of rainfall *etc.*
   a. *Onset of monsoon:* More than 70 per cent of rainfall is received during June to September in the country (Table 11). The summer monsoon normally arrives over South Kerala on 1st June. Thereafter, it continues to advance northwards and westwards and reaches Rajasthan by mid July. The actual date of monsoon onset over a place in a year undergoes variation of 8 days (Probability being 67%). Estimation of onset of monsoon is important for taking up dry sowing (Table 12).
   b. *Withdrawal of monsoon:* The normal monsoon withdrawal from western parts of Rajasthan commences by 1st September and the withdrawal is completed by October 15th then north-east monsoon sets in.

Table 11: Seasonal distribution of annual rainfall in India

Source and Duration	Seasonal distribution (%)
Summer monsoon (March – May)	10.4
South-west monsoon (June – September)	73.7
North-east monsoon (October – December)	13.3
Winter monsoon (January – February)	2.6

   c. *Rainfall distribution:* Long term studies on the behaviour of rainfall distribution shows that the availability of rain is confined to a period marked by a beginning and end. A small departure from the normal is enough to trigger drought like conditions.
   d. *Length of growing season (LGS):* LGS is the period during a year when precipitation exceeds half the potential evapotraspiration plus a period required to evapotraspiration an assumed 100 mm of water stored from precipitation in the soil profile (FAO definition). In dry farming, LGS depends on duration of rainy season and moisture retentive characteristics of soil. LGS is a period between seed germinating rains to end of season when profile moisture is reduced to a AE/PE to 0.5.
   e. *Dependability of rainfall:* Another important character of rainfall is its dependability. The measure used for describing degree of variability is the coefficient of variability {CV (%) = [σ/mean] × 100 }. The higher the CV, the higher is the variability and less is the dependability. In west of Surat to Delhi line, CV exceeds 30% and rise to 60 to 80% in west

of Rajasthan and Kutch. Elsewhere CV is 20 to 30% with lower values along the west coast and with east Assam. Monthly rainfall variability in the rainiest months (July and August) is as high as 40 to 50 per cent. CV further increases in September and October, the uncertainly becomes nearly complete. If these aspects are not taken note in crop planning risk of uncertainly with consequent failures is great.

Table 12: Effect of mean annual rainfall on coefficient of variation and probability of occurrence of deficit rainfall

Station	Mean annual rainfall (mm)	Annual rainfall cv (%)	Probability of occurrence of deficit rainfall (< 75% of normal)
Jodhpur	369	55	51
Anantpur	568	30	38
Hyderabad	767	29	31
Varanasi	1026	25	25
Ranchi	1434	21	20
Shillong	2415	15	7

The coefficient of variation in its occurrence, increases with a fall in annual precipitation (Table 12).

II. Recurrence of rainfall events

In the rainfed agriculture the question is how much rainfall can be expected at least in the time interval in three out of four years or four out of five years. This can be known by using ranking order method (as described by Doorenbos and Pruitt (1984) or Weibull's method). The main assumption in this method is that the rainfall is more or less normally distributed.

The return periods according to weibull's formula are calculated as under and arranged in a table (Table 13)

$$P = \left\{ \frac{m}{n+1} \right\} \times 100$$

Where, P = Plotting position in per cent chance, m = rank number when the data are arranged in descending order and n = Total number of years events weeks for which data is available

Table 13: Recurrence of rainfall events

Rainfall (P) order	Rainfall (m)	Plotting (Descending order) position/probability%
1	Highest	–
2	–	–
3	–	–
–	–	–
n	Lowest	–

If a probability level of 50 is required, it can be calculate as under. 30 = 100 (m)/(30 + 1) when n = 30 years/event

(***Note*:** If the quotient is 0.5 and above, it should be rounded of to next number).

In general, 75% probability is used in agriculture as dependable rainfall.

The information can be depicted in a graph using total amount of rainfall on Y-axis and probability recurrence interval on X-axis and useful conclusions could be drawn.

III. Estimation of rainfall variability based on Statistical predictions

For the purpose of analysing the variation of rainfall for a place over a period the data were subjected to analysis through the following statistical techniques:

1. Descriptive statistics
2. Trend analysis of rainfall for two meteorological stations
3. Markov Chain model to model dry, wet and rainy spells

1. Descriptive statistics

Descriptive statistics are brief *descriptive* coefficients that summarize a given data set, which can be either a representation of the entire or a sample of a population. *Descriptive statistics* are broken down into measures of central tendency and measures of variability (spread).

Maximum amount of rainfall, Minimum amount of rainfall, Mean, Standard Deviation, Coefficient of Variation and Skewness are the descriptive statistics computed for the study of rainfall variations over different seasons for GKVK meteorological station over period of 36 years from 1983 to 2018 (Table 14).

The annual rainfall of Bengaluru was found to be 928.81 mm for the given study period. The maximum amount of rainfall was found to be 1374.40 mm and the minimum amount of rainfall was found to be 528.00 mm for 36 years. The coefficient of variation was found to be 401.82 per cent. The annual total rainfall of Mysuru was 681.65 mm for the given study period. Mysuru witnessed the maximum amount of rainfall of 1085.90 mm and the minimum amount of total annual rainfall of 235.30 mm. The coefficient of variation was 371.29 per cent.

Table 14: Descriptive Statistics of Annual Total Rainfall (mm) for Bengaluru and Mysuru Districts from 1983–2018

District	Minimum	Maximum	Mean	SD	CV (%)	Skewness
Bengaluru	528.00	1374.40	928.81	231.14	401.82	0.19
Mysuru	235.30	1085.90	681.65	183.58	371.29	0.10

Source: Raaga, 2019

The total average rainfall for 36 years for three seasons (Pre-Monsoon, Monsoon and Post-Monsoon) of Beangaluru was 174.19 mm, 505.85 mm and 232.77 mm, respectively (Table 15). The coefficient of variation was minimum during Post-Monsoon season (166.93%). The total average rainfall of Mysuru for 36 years for three seasons (Pre-Monsoon, Monsoon and Post-Monsoon) was 175.38 mm, 289.46 mm and 204.82 mm, respectively. The coefficient of variation for Mysuru was found to be highest during Pre-Monsoon (255.31%) and was low during the Monsoon season (172.92%).

Table 15: Descriptive Statistics for Total Seasonal Rainfall (mm) for Bengaluru and Mysuru Districts from 1983–2018

District	Seasons	Min	Max	Mean	SD	CV (%)	Skewness
	Pre-Monsoon	52.60	328.60	174.19	67.80	256.89	0.46
Bengaluru	Monsoon	252.50	835.50	505.85	147.02	344.06	0.08
	Post-Monsoon	52.60	693.10	232.77	139.43	166.93	1.75
	Pre-Monsoon	32.90	365.20	175.38	68.69	255.31	0.28
Mysuru	Monsoon	84.50	514.10	289.46	98.51	293.81	0.24
	Post-Monsoon	36.90	521.20	204.82	118.44	172.92	0.72

Source: Raaga, 2019

2. Trend analysis of rainfall

Analysis of trend fits a particular type of trend line or curve to a time series data. In general, the magnitude of trend in a time series is determined either using parametric test *i.e.,* regression analysis or non-parametric method known as Sen's estimator method. Analysis was carried out using Excel, XLStat, SPSS and R-software which generates a time series plot that shows the original data and the fitted trend line.

The general linear trend equation is given below,

$$Y_x = a + bx + e$$

where,

Y_x = Trend values at time x.

a = Intercept parameter.

b = Slope parameter.

e = Error term.

x = Time period.

Sen's estimator

Sen's estimator has been widely used for determining the magnitude of trend in hydrometeorological time series. In this method, the slopes (T_i) of all data pairs are first calculated by

$$T_i = \frac{x_j - x_k}{j - k} \text{ for } i = 1, 2, 3, \cdots, N$$

where x_j and x_k are data values at time j and k ($j > k$) respectively. The median of these N values of T_i *is* Sen's estimator of slope which is calculated as

$$\beta = \begin{Bmatrix} T_{\frac{N+1}{2}} & N \text{ is odd} \\ \frac{1}{2}\left(T_{\frac{N}{2}} + T_{\frac{N+2}{2}}\right) & N \text{ is even} \end{Bmatrix}$$

A positive value of β indicates an upward (increasing) trend and a negative value indicate a downward (decreasing) trend in the time series. The trend line for all the three seasons and annually were computed for the two meteorological stations and compared with a non-parametric Mann-Kendall trend test.

Mann-Kendall trend test

The non-parametric Mann-Kendall trend test was used to determine the rainfall trend. Significance was tested stating that there is no trend in the series as null hypothesis (H_0) against the alternative hypothesis (H_1) that there is a negative, non-null or positive trend. It was tested at 95 per cent significance level. Since rainfall pattern shows the trend, the appropriate non-parametric test will be Mann- Kendall test. It is appropriate because its statistic is based on the sign of differences, not directly based on values of random variable and it is also less affected by the fluctuations. Monotonic trend in any time series data can be detected by the application of Mann- Kendall (non-parametric) test.

The test statistic S is calculated using the formula

$$S = \sum_{k=1}^{n-1} \sum_{j=k+1}^{n} \operatorname{sgn}(x_j - x_k)$$

$$\operatorname{sgn}(x_j - x_k) = \begin{Bmatrix} +1 & \text{if } x_j - x_k > 0 \\ 0 & \text{if } x_j - x_k = 0 \\ -1 & \text{if } x_j - x_k < 0 \end{Bmatrix}$$

where n is the number of observed data series, x_j and x_k are the values in periods j and k respectively, $j > k$. For $n \geq 10$, the sampling distribution of S is given by

$$z = \begin{cases} \dfrac{s-1}{\sqrt{\text{Var}(s)}} & \text{if } s > 0 \\ 0 & \text{if } s = 0 \\ \dfrac{s+1}{\sqrt{\text{Var}(s)}} & \text{if } s < 0 \end{cases}$$

where Var(S) is determined as

$$\text{Var}(s) = \frac{1}{18}\left[n(n-1)(2n+5) - \sum_{p=1}^{q} t_p(t_p - 1)(2t_p + 5) \right]$$

Trend analysis Bengaluru and Mysuru during the period of 36 years from 1983 to 2018 by use of non parametric test (Mann-Kendall) shows no trend since Kendall's trend test statistic value has the p value more than the 0.05 (level of significance) value (Table 16).

Table 16: Estimated Trend values in Annual Rainfall for Bengaluru and Mysuru districts

District	Intercept (S.E)	Slope (S.E)	R^2	Kendall's tau value	Sen's Slope	p-value
Bengaluru	929.294 (79.831)	–0.260 (3.763)	0.000	0.000	0.061	1.000
Mysuru	737.673 (62.441)	– 3.028 (2.943)	0.030	–0.095	–2.577	0.421

Source: Raaga, 2019

MK test for trend during different Nakshatra periods for meghadhipathy based classification

Indians have been using almanacs (Panchang) from Vedang Jyotish period (1,400–1,300 BC) (Sivaprakasam & Kanakasabai, 2009). Its use is noticed in the field of agriculture as they give the auspicious dates for land preparation, sowing, harvesting as well as to predict annual rainfall. Although there are many modern scientific methodologies for rainfall prediction, they have their own set of drawbacks. The new techniques of rainfall forecast can give the predictions only a few days prior to the occurrence of rains. Traditional knowledge is wisdom developed over many generations. Many farmers still believe in traditional methods for prediction rather than a scientific approach. Due to social and religious norms, they believe that scientific methods are harmful or not sacred. Also, native rural people are not much connected to the modern world due to superstitions. They believe in indigenous knowledge because their predictions work accurately according to them. In order to have better technology reach to them, its blending with traditional knowledge/ belief is required.

The daily rainfall data for 28 years from 1989–2017 was collected and used for the purpose of comparing almanac predictions with actual rainfall. The study was conducted purely based on secondary data. The required data was collected from Karnataka State Natural Disaster Monitoring Centre (KSNDMC), Bengaluru. The Panchang is modelled based on Hindu system of New Year which starts from the tithi Chaitra Shukla Padyami, which usually falls in the month of April and is known by the name Ugadi. This is believed to be the time when Lord Brahma, the Creator among the Hindu Trinity, started the conception of this Universe. This is a principal festivity in India. The year ends with the tithi Phalguna Krishna Paksha (Bahula) Amavasya, which normally occurs during the month of March. Based on this 1st April of the year to 31st March of next year was considered as one year.

Example: 1st April 1989 to 31st March 1990 is considered as one year, denoted as 1989–1990.

The day on which Sun enters Aridra Nakshatra will be considered as meghadhipathy for the year.

There was a strong belief that in each of the year, a minimum amount of rainfall was obtained during the assured period. During the study period, maximum rainfall was observed in Punarvasu Nakshatra followed by Pushya Nakshatra. The coefficient of variation is below 60 per cent from Rohini (May 25–June 7) to Hasta (September 27–October 10) indicating that in these ten Nakshatra periods, on an average rainfall observed with less variability which in turn indicates that during this Nakshatra period assured quantum of rainfall is received. The average rainfall received during different Nakshatra periods is depicted in Fig. 6 and Table 17.

Table 17: Descriptive statistics of rainfall during different stars and duration of the stars in the Gregorian calendar

Star	Period	Duration	Min.	Max.	Mean	SD	CV (%)
Ashwini	Apr 14–Apr 26	13	19.077	328.701	122.768	71.617	58.336
Bharani	Apr 27–May 10	14	36.013	475.490	154.503	90.916	58.844
Krithika	May 11–May 24	14	28.370	506.963	198.917	119.822	60.237
Rohini	May 25–Jun 7	14	60.860	880.590	384.869	200.983	52.221
Mrigashira	Jun 8–Jun 21	14	366.467	1151.203	567.563	164.792	29.035
Aridra	Jun 22–Jul 5	14	87.380	1414.237	583.562	297.926	51.053
Punarvasu	Jul 6–Jul 19	14	226.657	1255.013	665.617	229.994	34.553
Pushya	Jul 20–Aug 2	14	183.800	1252.057	663.951	244.104	36.765
Ashlesha	Aug 3–Aug 16	14	220.367	1045.107	572.892	200.358	34.973

(Contd.)

Star	Period	Duration	Min.	Max.	Mean	SD	CV (%)
Magha	Aug 17–Aug 30	14	222.080	834.020	498.233	175.755	35.276
Hubba	Aug 31–Sep 12	13	77.780	804.750	389.193	180.627	46.411
Uttara	Sep 13–Sep 26	14	58.160	812.440	391.880	217.551	55.515
Hasta	Sep 27–Oct 10	14	122.967	935.853	479.247	225.142	46.978
Chitta	Oct 11–Oct 23	13	65.853	801.817	324.912	203.486	62.628
Swathi	Oct 24–Nov 5	13	22.520	458.210	221.763	144.201	65.025
Vishakha	Nov 6–Nov 19	14	8.870	821.797	142.882	179.170	125.398
Anuradha	Nov 20–Dec 2	13	0.040	265.643	53.909	62.260	115.491
Jeshtha	Dec 3–Dec 15	13	0.000	247.803	29.799	58.011	194.677
Moola	Dec 16–Dec 28	13	0.000	85.880	10.120	20.336	200.944
Poorvashadha	Dec 29–Jan 10	13	0.000	95.937	10.048	20.533	204.343
Uttarashadha	Jan 11–Jan 23	13	0.000	61.540	8.490	17.060	200.953
Shravana	Jan 23–Feb 5	13	0.000	42.313	3.174	8.398	264.628
Dhanishtha	Feb 6–Feb 18	13	0.000	86.297	9.527	22.292	234.001
Shatabhisha	Feb 19–Mar 3	13	0.000	72.843	8.173	16.303	199.467
Poorvabhadra	Mar 4–Mar 17	14	0.000	180.920	31.311	49.488	158.053
Uttarabhadra	Mar 18–Mar 30	13	0.000	547.473	34.739	102.210	294.226
Revathi	Mar 31–Apr 13	14	2.630	254.590	79.413	61.957	78.019

Source: Aparna Bhide, 2019

Table 18: MK test to see the trend in rainfall during different Nakshatras

Star	Z	Q	p–value
Ashwini	1.640	2.705	0.101
Bharani	0.454	0.718	0.650
Krithika	0.612	1.585	0.540
Rohini	–0.494	–1.597	0.621
Mrigashira	–1.679	–5.613	0.093
Aridra	0.533	4.773	0.594
Punarvasu	–1.482	–7.572	0.138
Pushya	–0.533	–3.104	0.594
Ashlesha	–0.375	–1.933	0.707
Magha	0.138	0.556	0.890
Hubba	1.679	7.728	0.093
Uttara	0.889	5.478	0.374
Hasta	–1.521	–10.155	0.128
Chitta	–1.600	–8.205	0.110
Swathi	0.059	0.113	0.953

Star	Z	Q	p–value
Vishakha	–0.929	–1.894	0.353
Anuradha	0.652	0.265	0.514
Jeshtha	1.445	0.240	0.149
Moola	0.219	0.000	0.827
Poorvashadha	0.000	0.000	1.000
Uttarashadha	0.972	0.001	0.331
Shravana	1.878	0.014	0.060
Dhanishtha	0.693	0.009	0.489
Shatabhisha	1.629	0.025	0.103
Poorvabhadra	2.038	0.527	0.042
Uttarabhadra	0.790	0.120	0.429
Revathi	0.000	–0.050	1.000

Source: Aparna Bhide, 2019

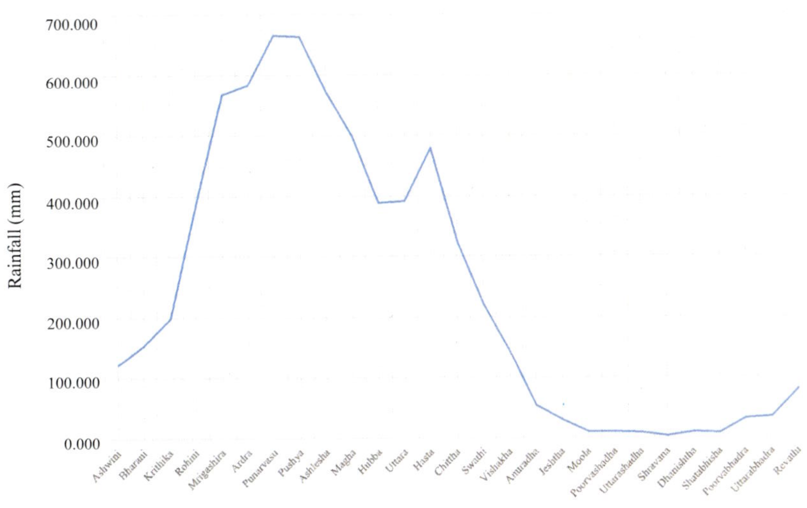

Source: Aparna Bhide, 2019

Fig. 6. Nakshatra wise average rainfall (mm) during 1989–2017

To analyse whether the rainfall pattern during Nakshatra periods follows any trend, Mann-Kendall (MK) trend test was employed. The magnitude of the trend was estimated using Sen's slope estimator. The results of the test are given in Table 18 which indicate that there was no significant trend followed except for Poorvabhadra Nakshatra. Only Poorvabhadra Nakshatra showed a significant positive trend of the magnitude 0.527.

3. Markov Chain model to model dry, wet and rainy spells

Markov chain probability model has been extensively used to find the long term frequency behaviour of wet and dry weather spells as well as for computation of probability or occurrence of daily precipitation. Its practical Utility is in agricultural planning under dryland ecosystem.

Markov Model is most powerful tool for analyzing complex system. Analyses yield results for both time dependent evolution of the system and steady state of the system. It also yields a variety of useful performance measure describing the operation of the system.

Past weather events exerts an influence on the future weather. The notion of persistence in the sequence of weather event is very important. Markovian models have been applied to climatologically processes exhibiting sequential patterning. Most of the studies dealt exclusively with the occurrence of wet and dry days. The sequence used to estimate the model is called training sequence and the sequence, which is used to evaluate, or to test the performance of the model is called test sequence.

The daily rainfall data of 36 years from 1983–2018 were expressed in Standard Meteorological Weeks (SMW) for analysis. As per the IMD classification, a week was classified as dry, wet and rainy, if the total rainfall is less than 10 mm, between 10 mm to 20 mm and above 20 mm, respectively. This classification of rainfall was considered as the possible states of dry, wet and rainy for the application of three state Markov process to study the pattern of occurrence of rainfall in Bengaluru and Mysuru meteorological stations during Pre-monsoon, Monsoon and Post-Monsoon periods. Further, the expected length of dry, wet and rainy spells and combination of these as weather cycle have been derived for each season based on weekly rainfall data.

A three state Markov chain model by considering a 3x3 transition probability matrix defined as shown in Table 19.

Table 19: 3 × 3 Transition probability matrix

Transition Probability Matrix		**Current Week**		
		Dry(d)	**Wet(w)**	**Rainy(r)**
Previous week	Dry(d)	P_{dd}	P_{dw}	P_{dr}
	Wet(w)	P_{wd}	P_{ww}	P_{wr}
	Rainy(r)	P_{rd}	P_{rw}	P_{rr}

Source: Raaga, 2019

where,

P_{dd} = Probability of a dry week preceded by a dry week,

P_{dw} = Probability of a wet week preceded by a dry week,

P_{dr} = Probability of a rainy week preceded by a dry week and so on.

For three-state Markov chain model, the values of P_{ij} (d, w, r) can be estimated from the 3×3 order observed frequency as shown in Table 20:

Table 20: 3 × 3 Observed frequency matrix

Transition Probability Matrix		**Current Week**			**Total**
		Dry(d)	**Wet(w)**	**Rainy(r)**	
Previous week	Dry(d)	n_{dd}	n_{dw}	n_{dr}	n_d
	Wet(w)	n_{wd}	n_{ww}	n_{wr}	n_w
	Rainy(r)	n_{rd}	n_{rw}	n_{rr}	n_r

Source: Raaga, 2019

where,

$n_{ij,}$ $i, j = (d, w, r)$ is observed frequency *i.e.*

n_{dd}: Number of dry weeks preceded by dry weeks

n_{dw}: Number of wet weeks preceded by dry weeks

n_{dr}: Number of rainy weeks preceded by dry weeks and so on.

where,

$n_d = n_{dd} + n_{dw} + n_{dr}$, Total number of dry weeks

$n_w = n_{wd} + n_{ww} + n_{wr}$, Total number of wet weeks

$n_r = n_{rd} + n_{rw} + n_{rr}$, Total number of rainy weeks

The maximum likelihood estimators of $P_{i,j}$ (d, w, r) are given by,

$$\widehat{P_{ij}} = \frac{n_{ij}}{\sum_{j=d}^{r} d^{n_{ij}}}$$

Test of Goodness of fit

To ascertain the basic assumption of the Markov chain model, one-week dependence could be considered for explaining the behaviour of rainfall, traditional Chi-square test and the test suggested by Wang and Martiz (1990) was applied. The hypothesis was stated as follows:

H_0: The rainfall occurrence on successive weeks was independent
H_1: The rainfall occurrence on successive days was not independent

The traditional Chi-square test statistic is

$$x^2 = \sum_{i=1}^{n}\left[\frac{(O_i - E_i)^2}{E_i}\right]$$

where,

O_i (i = 1, 2..., n) are observed frequencies,

E_i(i =1, 2..., n) are corresponding expected frequencies.

i = number of observations (i = 1, 2,3…)

The critical region for testing null hypothesis is $X_c^2 \geq X_{(n-1,\alpha)}^2$ where X_c^2 the calculated value of Chi-square and $X_{(n-1,\alpha)}^2$ is tabulated value of Chi-square for (n – 1) degree of freedom and a level of significance.

During 1990, Wang and Maritz observed that WS test statistic for independence of three-state Markov Chain is better than the traditional-test method. According to Wang and Maritz, the test statistic, for testing the above null hypothesis, is

$$WS = \frac{S_a + S_b - 1}{\sqrt{V(S_a + S_b - 1)}} \sim N(0,\ 1)$$

where,

$$S_a = P_{dd} + P_{ww} + P_{rr}$$

$$S_b = P_{rd}P_{dr} + P_{wr}P_{rw} + P_{dw}P_{wd} - P_{dd}P_{ww} - P_{ww}P_{rr} - P_{rr}P_{dd}$$

$$V(S_a + S_b - 1) = (2p_1P_2P_3)\left[\frac{1}{n_d n_w} + \frac{1}{n_w n_r}\frac{1}{n_r n_d}\right]$$

where, WS= Wilcoxson Signed Rank Test

$n_{d.}$, $n_{w.}$ and $n_{r.}$ are the total number of dry, wet and rainy weeks and p_1, p_2 and p_3 are the stationary probabilities calculated as follows,

$$p_1 = \left[(1+p) + \frac{(1+s)p}{q}\right]^{-1}$$

$$p_2 = \left[r + \frac{ps}{q}\right]p_1$$

$$p_3 = \left[\frac{p}{q}\right]p_1$$

where: p, q, r and s can be calculated using the formula

$$p = \left[p_{dr} + \frac{p_{wr}(1 - p_{dd})}{p_{wd}}\right]\left(\frac{1}{1 - p_{rr}}\right)$$

$$q = 1 + \left[\frac{p_{wr} P_{rd})}{p_{wd}(1 - p_{rr})} \right]$$

$$r = \left[\frac{p_{dw}}{(1 - p_{ww})} \right]$$

$$s = \left[\frac{p_{rw}}{(1 - p_{ww})} \right]$$

If the test statistic calculated is greater than or equal to critical value of SND at 'a' level of significance, we reject the null hypothesis.

Calculation of Long Run (Equilibrium) Probabilities

The estimates of the long run equilibrium probabilities for each of three periods were calculated as given below:

$$\pi_1 = p_1 P_{dd} + p_2 P_{wd} + p_3 P_{rd}$$

$$\pi_2 = p_1 P_{dw} + p_2 P_{ww} + p_3 P_{rw}$$

$$\pi_3 = p_1 P_{dr} + p_2 P_{wr} + p_3 P_{rr}$$

where, π_i's (i = 1, 2, 3) are the probabilities of week being a dry, wet and rainy, respectively.

Expected Length of Different Spells of Seasons and Weather Cycle

The following formulas were used to calculate expected length of different spells and weather cycle:

$$E(D) = \frac{1}{(1 - P_{dd})}$$

$$E(W) = \frac{1}{(1 - P_{ww})}$$

$$E(R) = \frac{1}{(1 - P_{rr})}$$

$$E(WC) = E(D) + E(W) = E(R)$$

which gives the expected duration of dry, wet, rainy weeks and expected length of weather cycle.

Before applying Three-state Markov Chain Model, the hypothesis of dependency of rainfall occurrences on successive weeks were tested separately for three different seasonal periods for two meteorological stations using traditional Chi-Square statistic, WS statistic and the statistic with their p-values for Bengaluru district are presented in Table 21. For all three periods based on weekly rainfall data, it was clear that according to Chi-Square and WS test statistic the null hypothesis of independence of rainfall on consecutive week was rejected at 5 per cent level of significance. This indicates that current week rainfall depends on preceding week rainfall.

Table 21: Estimated Chi-Square statistic and WS statistic associated with their p-values for Bengaluru district

District	Test Statistic	Pre-Monsoon	Monsoon	Post-Monsoon
Bengaluru	Chi-Square (p–values)	51.71 ($p < 0.00001$)	19.19 ($p = 0.00072$)	51.13 ($p < 0.00001$)
	WS- Statistic (p–values)	3.29 ($p = 0.001002$)	8.34 ($p < 0.00001$)	15.85 ($p < 0.00001$)

Source: Raaga, 2019

As the current week rainfall depends on preceding week rainfall, three-state Markov Chain model was used to study the behavior of rainfall. The estimated transition probability matrices for weekly rainfall data for Bengaluru are reported in Table 22 (Pre-Monsoon), Table 23 (Monsoon) and Table 24 (Post-Monsoon), respectively. Estimated transition probability matrices reported that the chances of the dry week followed by the dry week for three seasons was found to be 80 per cent, 44 per cent and 71 per cent, respectively for Bengaluru district. The chances of the wet week preceding the wet week for three different seasons was observed to be 14 per cent, 26 per cent and 11 per cent, respectively. Similarly, it was observed that the chances of the rainy week preceding the rainy week were found to be 40 per cent, 48 per cent and 53 per cent for three different seasons, respectively.

Table 22: Estimated Transition probabilities for Pre-Monsoon season for Bengaluru district

	Transitional Probabilities		
	Dry	**Wet**	**Rainy**
Dry	0.80	0.06	0.14
Wet	0.50	0.14	0.36
Rainy	0.48	0.12	0.40

Source: Raaga, 2019

Table 23: Estimated Transition probabilities for Monsoon season for Bengaluru district

	Transitional Probabilities		
	Dry	**Wet**	**Rainy**
Dry	0.44	0.22	0.34
Wet	0.34	0.26	0.40
Rainy	0.38	0.14	0.48

Source: Raaga, 2019

Table 24: Estimated Transition probabilities for Post-Monsoon season for Bengaluru district

	Transitional Probabilities		
	Dry	**Wet**	**Rainy**
Dry	0.71	0.10	0.19
Wet	0.5	0.11	0.3
Rainy	0.35	0.12	0.53

Source: Raaga, 2019

The three-state Markov chain technique adopted enables easy computation of the equilibrium state probabilities p_1, p_2 and p_3 of a week being dry, wet and rainy, respectively. The expected length of different spells, the weather cycles and the total numbers of days in each period over 36 years for Bengaluru district as presented in Table 25. It was observed that the expected lengths of dry, wet and rainy days are 5 weeks, 1 week and 2 weeks, respectively with a weather cycle of 8 weeks for the Pre-monsoon season; 2 weeks, 1 week and 2 weeks, respectively with a weather cycle of 5 weeks for the Monsoon season and 4 weeks, 1 week and 2 weeks, respectively with a weather cycle of 7 weeks Post-Monsoon for season for Bengaluru district.

Table 25: Estimated equilibrium state probabilities, the expected length of different spells, the weather cycle and the total number of weeks in each period for Bengaluru district

Bengaluru	π_1	π_2	π_3	**Expected Length of Spells**				**Total No. of weeks**
				Dry Run	**Wet Run**	**Rainy Run**	**Weather Cycle**	
Pre-Monsoon	0.71	0.07	0.22	5.00	1.00	2.00	8.00	13
Monsoon	0.62	0.12	0.26	2.00	1.00	2.00	5.00	17
Post-Monsoon	0.57	0.21	0.22	4.00	1.00	2.00	7.00	13

Source: Raaga, 2019

It was evident from Table 25, that the chances of the week being dry, wet and rainy during Pre-Monsoon season was 71 per cent, 7 per cent and 22 per cent, respectively. While the week being dry, wet and rainy during Monsoon season was found to be 62 per cent,12 per cent and 26 per cent and for Post-monsoon season, it was observed that 57 per cent, 21 per cent and 22 per cent of weeks were dry, wet and rainy, respectively.

It was clear from the transitional probabilities that more number dry weeks preceding dry weeks was observed during Pre-Monsoon season, more wet weeks preceding wet weeks were observed during Monsoon season and more rainy weeks preceding rainy weeks was during Post-Monsoon season for Bengaluru district. From Table 25 Post-Monsoon season had more rainy weeks rather than the Monsoon season which indicates that there was a shift in rainy weeks. From the estimated equilibrium probabilities more chances of dry weeks were observed during Pre-monsoon season, more wet weeks were during Post-Monsoon season and rainy during Monsoon season. During Monsoon season it was observed that 26 per cent of weeks were rainy weeks.

Chapter 5

Classification of Climate

Climatic classification is the method of identifying climatologically homogenous regions. The systematic study of climate and its analysis and processing, which is referred as climatology plays an important role in national weather, planning activities and decision making in human activities. Iso climatic regions are those regions which have same climatic conditions.

Specific objectives of climatic classification

1. To identify the climates for transfer of production technology developed at particular center.
2. To provide adequate indices of moisture availability for successful crop production in drylands.
3. To quantify the risk levels associated with production constrains to agriculture production.

Principles of climatic classification

The basic principles under which a climate can be classified are listed as follows:

1. It should provide a concise description of climatic types of the truly active factors (moisture, heat, wind, sunshine, pressure changes *etc.*)
2. It should provide a system ranging from microclimate to macroclimate.
3. It should be made from adequate records with respect to periods and distribution.
4. It should consider the systematic relationship between climatic factors and pattern of vegetation.
5. It should consider the factors like human health and comfort (style of clothing, housing *etc.*).

The Climatic classification by different methods are as follows:

1. Lang's classification

Lang (1920) given the climatic classification based annual precipitation (P) and mean annual temperature (T) ratio. On the basis of P-T ratio the climatic classification as follows:

P–T ratio	Climate
10–40	Arid
40–160	Humid
160 & above	Wet

Where,

P = annual precipitation (mm)

T = mean annual temperature (°C)

2. De Martonne's classification

De Martonne (1926) given the following index to limit different vegetational zones of the earth. This is slight modification of Lang's rain factor

$$I = \frac{\text{Annual precipitation (mm)}}{\text{Mean annual temperature (°C)} + 10}$$

I_{DM} value	Climatic condition
$I_{DM} < 5$	Hyper- arid
$5 < I_{DM} < 10$	Arid
$10 < I_{DM} < 20$	Semi- arid
$20 < I_{DM} < 24$	Mediterranean
$24 < I_{DM} < 28$	Sub-humid
$28 < I_{DM} < 35$	Humid
$35 < I_{DM} < 55$	Very humid
$55 < I_{DM}$	Extremely humid

3. Koppen's classification

Koppen (1936) grouped the world climate into the following groups:

1. Tropical rainy climate (A) with temperature of the coolest month exceeding 18°C. This is subdivided into three sub-groups, *viz.,* no dry season (f), dry season in summer (s), and dry season in winter (w).
2. Dry climate (B) with sub-divisions of semi-arid or steppe (s), and arid or desert (w) (Tropical deserts and steppe have average annual temperatures exceeding 18°C).
3. Warm temperature rainy climate (C) with average temperature of the coldest month being less than 18°C but greater than –3°C. This is again sub-divided into no dry season in summer (s), and dry season in winter (w).
4. Cold snow climate (D) with average temperature of the coldest month < –3°C but with the average temperature of the warmest month greater than 10°C. This is also sub-divided into 3 sub-groups as in the previous case.

5. Polar climate (E) with average temperature of even the warmest month being less than 10 °C. This climate covers Tundra and ice-cap regions.

Boundaries of climate for maximum rainfall in summer and winter as proposed by Koppen, 1936

Rainfall (inches)	Boundary between dry and tree climates	Boundary between steppe and desert climates
Maximum in summer	$r = 0.44\,t - 3$	$r = \frac{0.44t - 3}{2}$

4. Gaussen's classification

A dry month is defined as a month in which the total precipitation (P) expressed in mm is equal to or less than twice the mean temperature (T) of the month expressed in degree centigrade.

$P < 2T$

An ombrothermic diagram drawn with these parameters indicates the duration and severity of dry months. The classification is based on xerothermic index which takes into account factors of temperature, precipitation, number of days of rain, atmospheric humidity, mist and dew.

The climate is classified into warm, warm temperate or temperate, when mean temperature is greater than zero and cold temperate if negative mean temperature occurs at certain period of the year. If the mean temperatures is negative for all months of the year is classified as glacial climate of the region.

When the xerothermic index is over 300, a region is termed a desert, and when it is between 200 and 300 the region is classified as subdesert. Mediterranean climate is classified for xerothermic index values ranging from 0 to 200, with a number of subdivisions depending on the range of the index values. Axeric climates have a xerothermic index of 0. Since rainfall is adequate for vegetation in this case, the factor for delimitation of the bioclimates is mean temperature (t), as indicated below.

$t > 20°C$ Equatorial

$15° < t < 20°C$ Subequatorial

$10° < t < 15°C$ Warm temperate

$0° < t < 10°C$ Temperate

Similar subclassification of cold and cold temperate climates is done on the basis of the combined duration of frosts and dry seasons in the case of desert, subdesert, and subaxeric climates, and the length of the frosty season in respect of cold axeric climates. Detailed study of xerothermic indices in India and climatic classification by this method was done by Legris and Viart (1959) and Meher-Homji (1960).

5. Emberger's classification

Emberger (1955) gave the delimitation of various bio-climates by means of Pluviothermic quotient (Q) defined as

$$Q = \frac{100P}{(M-m)(M+m)}$$

Where, M is the mean maximum temperature (°C) in the hottest month, m is the mean minimum temperature (°C) in the coldest month, and P is annual rainfall (mm)

6. Thornthwaite classification

Koppen's climatic classification is lacking of rational basis for selecting temperature and precipitation values for different climatic zones. Thornthwaite improved the same by introducing water balance concept in this classification. The moisture index (I_m) is determined by the following relation:

Climate types	Moisture index (Im)
A-Per-humid	100 and above
B_4-Humid	80–100
B_3-Humid	60– 80
B_2-Humid	40–60
B_1-Humid	20–40
C_2-Moist sub-humid	0–20
C_1-Dry sub-humid	−20–0
D-Semiarid	−40 to −20
E-Arid	−60 to −40

Delinations of climate on the basis of moisture index (Thornthwaite, 1948)

$$I_m = I_h - 0.61a = \frac{100S - 60d}{n}$$

Where, $I_h = \frac{100S}{n}$ is the humidity index

and Aridity index $(I_a) = \frac{100d}{n}$

Where,

S – Water surplus
d – Water deficiency
n – Water need, and
I_m – The moisture index

In the moist climates if there is a dry season, one needs to know how is it dry and in the dry climates how wet the season is, if one occurs. In moist climates water deficiency may be large, moderate, small or non-existent. In dry climates the same is true for water surplus. Therefore, it is important to know whether a place is continuously wet or continuously dry or whether it is wet in one season and dry in another.

Thornthwaite classified the following sub-divisions in terms of the humidity and aridity indices as,

Moist climates (A, B, C_2)	**Aridity indices**
r – Little or no water deficiency	0–10
S_1 – Moderate summer water deficiency	10–20
W_1 – Moderate winter water deficiency	10–20
S_2 – Large summer water deficiency	> 20
W_2 – Large winter water deficiency	> 20

Dry climates (C_1, D, E)	**Humidity indices**
D – Little or no water surplus	0–16.7
S_1 – Moderate summer water surplus	16.7–33.3
W_1 – Moderate winter water surplus	16.7–33.3
W_2 – Large summer water surplus	> 33.3
S_2 – Large winter water surplus	> 33.3

For determination of aridity and humidity indices potential evapotranspiration is determined by Thornthwaite method. Potential evapotranspiration specifies it as an expression of day length as well as of temperature. Hence the potential evapotranspiration can be used as an index of thermal efficiency. It is not merely a growth index but expresses growth in terms of water that is needed for growth.

If seasonal variation in temperature is very little, a mean annual temperature of 23°C is taken as the boundary between megathermal and mesothermal climates in equatorial regions. Similarly seasonal variations in temperature, a mean annul temperature of 21.5°C is taken as the boundary in sub-tropical areas because reduced growth and low water need of winter are more than offset by an accelerated growth and increased water need in summer. For a station in equator with a mean temperature of 23°C during each month, the potential evapotranspiration is 114.0 cm. Thornthwaite index used for separating megathermal and mesothermal climates and other limits were taken for boundaries in an arithmetic progression with a common difference of 14 cm. Thornthwaite specified various limits for climatic type is as :

T-E index	Climate type	
	E' Frost	Microthermal
14.2 cm	D' Tundra	
28.5 cm	C_1'	
42.7 cm	C_2'	Mesothermal
57.0 cm	B_1'	
71.2 cm	B_2'	
85.5 cm	B_3'	
99.7 cm	B_4'	Megathermal
114.0 cm	A'	

7. Modified climatic classification of Thornthwaite and Mather (1955) The revised moisture index of Thornthwaite and Mather (1955) method can be written as

$$I_m = \frac{P - AE - PE + AE}{PE} \times 100$$

$$I_m = \frac{P - PE}{PE} \times 100$$

Where,

P = Precipitation (mm),

PE = Potential evapotranspiration (mm), and

AE = Actual potential evapotranspiration (mm

$$\text{Moisture Index (Im)} = \frac{P - PE}{PE} \times 100$$

Using Moisture Index (Im) the following classification were made

Im	100 and above	20 to 100	0 to 20	–33.3 to 0	–66.7 to –33.3	–100 to –66.7
Climate	Per humid	Humid	Moist sub humid	Dry sub humid	Semiarid	Arid

8. Troll's Classification (1965): Troll took temperature and humid months as parameters for distinguishing different climate. In this classification, tropical region is the one where mean monthly temperature of coolest month is more than 18^0C. Tropical region subdivided into five groups based on humid month. Humid month is one having mean rainfall exceeding the mean Potential evapo transpiration.

Classification of tropical region (Troll, 1965)

Symbol	Type of climate	Type of vegetation	Number of humid months
V1	Tropical rainy climate	Evergreen forest	12–9.5
V2	Tropical humid summer	Evergreen forest	9.5–7.0
V2a	Tropical humid winter	Half deciduous	9.5–7.0
V3	Tropical wet-dry	Rain forest dry wood	7.0–4.5
V4	Tropical dry climate	Thorny, succulent, wood	4.5–2.0
V4a	Tropical dry climate with humid		4.5–2.0
V5	Tropical semi-desert and desert climate		<2.0

Similar to tropical region, temperate and polar regions are subdivided. The main drawback of troll's classification is that the continuity of humid months is not considered.

9. Moisture Deficit Index

The ICAR under All India Coordinated Research Project on Dryland Agriculture of Indian Council of Agriculture Research adopted classification based on Moisture deficit index (MDI)

$$\text{Moisture Deficit Index (MDI)} = \frac{P - PET}{PET} \times 100$$

Where

P = Annual Precipitation (cm)

PE = Potential evapo-transpiration

Based on MDI the climate is divided into three regions as below:

Climate	Sub humid	Semiarid	Arid
Moisture Deficit Index (MDI)	0.0 to –33.3	−33.3 to −66.6	> −66.6

10. Hargreaves Method (1971)

This method is based on the degree of moisture deficit for agricultural production. According to this method:

$MAI = PD/PE$ = Amount of rainfall at 75% probability/Potential evapo-transpiration

Climate classification based on the moisture availability index (MAI) at 75% probability of rainfall

Range of MAI	Duration	Type of climate	Crop planning
0–0.33	All months	Very arid	Not suitable for rainfed crops
0.34 and above	1–2 month	Arid	Limited suitability for rainfed crops
0.34 and above	3–4 months	Semi arid	Limited suitability for rainfed crops
0.34 and above	5 or more consecutive months	Wet	Suitable for crops of 5 months or more duration

Hargreaves (1975) proposed the following moisture deficit classification for all types of climate

Range of MAI	Types of climate
0.00–0.33	Very deficit
0.34–0.67	Moderately deficit
0.68–1.00	Somewhat deficit
1.00–1.33	Adequate moisture
>1.34	Excessive moisture

The probability level as well as range of MAI appears to be very high. Different probability levels may be more appropriate for some crops under special condition.

11. Optimum Moisture Availability Index (*OMAI*)

This index was given by Sarkar and Biswas (1980) (Agro climate classification of India). According to this method:

$$\text{OMAI} = \frac{\text{Assumed rainfall at 50\% probability level}}{\text{Potential evapotranspiration}}$$

12. Papadakis Method (1975)

This classification is based on thermic and hydric indices.

Thermic scales takes into account:

- Average daily maximum temperature
- Average daily minimum temperature
- Average of the lowest temperature
- Length of frost free period

Hydric scale: It takes into account monthly precipitation (P), potential evapotranspiration (PET) and water stored in the soil (W) from the previous rains. For determining the hydric type of climate, mean monthly potential evapotranspiration (PET) can be determined by using mean maximum temperature and vapour pressure.

PET = 5.625 ($e_{ma} - e_d$)

Where, PET = Potential evapotranspiration in mm

e_{ma} = saturation vapour pressure (mb) corresponding to average daily maximum temperature

e_d = average vapour pressure of the month (mb)

Hydric scale: $P + W/PET$ = Precipitation + Water stored in the soil/Monthly potential evapotranspiration

On the basis of this, the following hydric types of climate are given:

Hydric Scale	Type of climate
< 0.25	Arid (a)
0.25–0.50	Dry (s)
0.50–0.75	Intermediate (e)
0.74–1.00	Intermediate humid (Y)
>1.00 (when P < PET)	Post humid (P)
1.00 –2.00 when $(P + W) - E < 100$ mm	Humid (h)
2.00 (when $(P + W) - \mathrm{E} > 100$ mm	Wet (W)

Based on thermic and hydric indices, the distribution of crops could be explained

13. Krishnan and Mukhtar Singh's Method (1972)

India was divided into different agroclimatic region based on moisture and thermal indices:

$$\text{Moisture Index (MI)} = \frac{P - PET}{PET} \times 100$$

Where,

P = Precipitation

PET = Potential evapotranspiration

$$\text{Thermal Index (TI)} = \frac{T_{\max} + T_{\min}}{2}$$

14. Sharma, Singh and Yadav's method (1978)

This method is based on the moisture index. Haryana was divided into seven agroclimatic regions. The moisture index is given below:

$$\text{Moisture Index (MI)} = \frac{P + I - PET}{PET} \times 100$$

Where,

P = Precipitation (cm)

I = Irrigation water (cm per unit area)

PET = Potential evapotranspiration

15. Mavi and Mahi's Method (1978)

In this method, agroclimatic regions of Punjab were based on weekly soil moisture index for summer season

$$\text{Soil moisture index (I)} = \frac{R + SM}{PE}$$

Where,

R = Rainfall at 25% probability level (mm)

SM = Soil moisture stored in the root zone (mm)

PE = Open pan evaporation (mm)

Chapter 6

Soil and Moisture Conservation

Soil is a living, dynamic ecosystem that nurtures plants. Soil is responsible for Sustaining plant and animal life below and above the surface, Regulation water and fertilizer/pesticide runoff, Storing and cycling nutrients. But, the natural soil balance was disturbed by man due to agricultural activities to get food. The intensification of agriculture leads to soil erosion, which leads to creation of unproductive soils.

In agriculture, soil erosion refers to the wearing away of a field's topsoil by the natural physical forces of water and wind or through forces associated with farming activities such as tillage. As a result of improper cultivation practices, deforestation and overgrazing aggravates the problem of soil erosion.

Globally soil erosion has reached a level that endangers the sustainable supply of food for the growing global population. It already threatens food production and ecosystem service delivery and therefore, there is a pressing need to address this threat. This is especially true in India where, in a total area of 328 million hectares (m ha), 121 m ha is undergoing soil degradation, 68% of which is attributed to water erosion. Water erosion rates range from 5 t ha^{-1} yr^{-1} in dense forest regions to rates in excess of 80 t ha^{-1} yr^{-1} where erosion is most severe, such as in the Shiwalik mountainous region. India's average soil loss has been estimated to be 15t ha^{-1} yr^{-1}; however, given the limited coverage of measurements, this should be treated with caution. In the light of the serious threat that soil erosion poses in the country, there is a pressing need for a systematic nationwide assessment of land degradation due to erosion using appropriate techniques. The area under ravine soil erosion is highest in UP to a tune of 12.30 lakh hectares and lowest in Maharashtra to a tune of 0.20 lakh hectares (Table 27).

Table 27: Area under Ravines in India (lakh hectares)

State	Area
Uttar Pradesh	12.30
Madhya Pradesh	6.83
Rajasthan	4.52
Gujarat	4.00
Punjab	1.20
West Bengal	1.04
Maharashtra	0.20

Effects of Soil Erosion

1. Erosion removes the top soil which is responsible for both physical support and nutrient supply to plants.
2. Loss of fertile soil due to soil erosion will result in reduced soil productivity.
3. Soil erosion and resultant deposition of soil may reduce the storage capacity of lakes and reservoirs and shortens their life.
4. Depth of drainage channels will be reduced, thus increasing over flows and flooding of the down streams. This can cause damage to crops and disaster to man made structures.
5. Replenishment of groundwater will be reduced due to increased runoff and reduced infiltration.
6. The soil may be fragmented due to the formation of gullies and the land value may be reduced.
7. Fertile lands may become unproductive by the deposition of eroded coarser materials.
8. Soil loss directly leads to the reduction in moisture retention and water holding capacity.
9. Detachment and removal of top soil result in requirement of more efforts for tillage operations.
10. Removal of top soil makes the survival of wild life more difficult as it destroys the natural habitat.

Types of Soil Erosion

The major types of soil erosion are caused by water and wind. Water is the major responsible factor causing soil erosion followed by wind and glaciers.

Water Erosion

It consists of two broad types namely geological and accelerated erosion. Geological erosion occurs in natural environment with no influence of rain and caused primarily by the action of water, wind, temperature variations and glaciers. Whereas accelerated erosion is mainly because of human activities like land preparation for growing crops, encroachment of land for construction of houses, industries *etc.*

The various types of water erosion caused by rainfall and other *Sources* of water are as follows

1. Splash erosion (otherwise referred as rain drop erosion)
2. Sheet erosion
3. Rill erosion
4. Gully erosion
5. Land slide erosion

Splash erosion

It results from soil splash by the falling raindrops. Rain drops having a velocity more than 50 kilometers per hour may splash to a height of two feet and spread of five feet. A good ground cover either by vegetation or mulching reduces the drop into sprays and increases the infiltration rate. Other factors influencing the splash erosion are intensity of rainfall, wind velocity and direction, soil type and slope of the land. The effect of splash erosion will be severe in slope lands.

Sheet erosion

It is characterized by uniform removal of soil in thin layers from slope lands. This form of erosion normally results from splash erosion and over flow on land surface. It is agriculturally an important type of erosion ending with the removal of valuable organic matter from soil.

Rill erosion

Sheet erosion occurs mostly in plain lands. In case of lands characterized by terrains or undulations, there will be accumulation of water in depressions and water will begin to flow the path offering least resistance, leading to formation of rills and removal of soil. Soil removal and transport will be more severe than sheet erosion and further, the removal will be aggravated under more intense storms.

Gully erosion

Advanced stage of rill erosion is called as gully erosion. Gullies occur in large extent than rill erosion and usually in the shape of V or U. The water flow through these gullies transports soil material.

Land slide erosion

Land slides are down slope movements of masses of soil and/or rock. They can cause some of the most spectacular damage attributable to earth movement or ground movement. Landslides include mud flows ("mud slides"), earth slumps, rockfalls, and other types of slope failures. Landslides can be fast or slow, wet or dry, small or large, shallow or deep, and reactivated or brand new. Triggering mechanisms can include rainstorms, landscape irrigation, broken pipelines, grading, inadequate surface drainage, erosion, earthquakes, and other natural phenomena and human activities. Fast-moving mud flows and debris flows are of particular concern in wildfire burn areas. Slope instability is a complex phenomenon that includes not only landslides but more subtle processes such as soil creep.

Based on various agro-climate and topographical situations, Universal soil loss equation (USLE) given by Wischemeir and Smith (1978) to estimate the soil loss by erosion.

A = R K L S C P

Where

A = Computed soil loss per unit area (t ha^{-1} $year^{-1}$)

R = Rainfall and Runoff factor, is the no. of erosion index units in a normal years rain. The erosion index is a measure of the erosion force of specific rainfall.

K = Soil erodibility factor, is the erosion rate per unit of erosion index for a specific soil in cultivated continuous fallow, on a 9% slope, 22 m long.

L = Slope-Length factor, is the ratio of soil loss from the field slope length to that from a 22m length on the same soil type and gradient.

S = Slope-gradient factor, is the ratio of soil loss from the field gradient to that from a 9% slope.

C = The crop management factor, is the ratio of soil loss from a field with specified crop with management to that from the fallow condition on which the factor K is evaluated.

P = Erosion-control practice factor (Support practice factor), is the ratio of soil loss with contouring, strip cropping or terracing to that with straight row farming, up and down slope.

Calculation

Table 28: Rainfall or runoff factor

Weather station	Upper Tier municipality Designation	R factor
Delhi	–	100
London	Countries of Lambton, Middlesex, and Oxford	100
Niagara	Regional Municipality of Niagara	90
Ridgetown	Municipality of Chatham-Kent	110
Toronto	City of Toronto, Regional Municipalities of Peel and York	90

Table 29: Soil erodibility Factor

Textural Class	K Factor tonnes/hectare (tons/acre)		
	Average OMC*	Less than 2% OMC	More than 2% OMC
Clay	0.49 (0.22)	0.54 (0.24)	0.47 (0.21)
Clay loam	0.67 (0.30)	0.74 (0.33)	0.63 (0.28)
Coarse sandy loam	0.16 (0.07)	–	0.16 (0.07)
Fine sand	0.18 (0.08)	0.20 (0.09)	0.13 (0.06)
Fine sandy loam	0.40 (0.18)	0.49 (0.22)	0.38 (0.17)
Heavy clay	0.38 (0.17)	0.43 (0.19)	0.34 (0.15)
Loam	0.67 (0.30)	0.76 (0.34)	0.58 (0.26)
Loamy fine sand	0.25 (0.11)	0.34 (0.15)	0.20 (0.09)
Loamy sand	0.09 (0.04)	0.11 (0.05)	0.09 (0.04)
Loamy very fine sand	0.87 (0.39)	0.99 (0.44)	0.56 (0.25)
Sand	0.04 (0.02)	0.07 (0.03)	0.02 (0.01)
Sandy clay loam	0.45 (0.20)	–	0.45 (0.20)
Sandy loam	0.29 (0.13)	0.31 (0.14)	0.27 (0.12)
Silt loam	0.85 (0.38)	0.92 (0.41)	0.83 (0.37)
Silty clay	0.58 (0.26)	0.61 (0.27)	0.58 (0.26)
Silty clay loam	0.72 (0.32)	0.79 (0.35)	0.67 (0.30)
Very fine sand	0.96 (0.43)	1.03 (0.46)	0.83 (0.37)
Very fine sandy loam	0.79 (0.35)	0.92 (0.41)	0.74 (0.33)

*Organic matter content

Table 30. LS Factor Calculation

Slope Length: m (ft)	Slope (%)	LS Factor
30.5 (100)	10	1.38
	8	1.00
	6	0.67
	5	0.54
	4	0.40
	3	0.30
	2	0.20
	1	0.13
	0	0.07
61 (200)	10	1.95
	8	1.41
	6	0.95
	5	0.76
	4	0.53
	3	0.39
	2	0.25
	1	0.16
	0	0.08
122 (400)	10	2.76
	8	1.99
	6	1.35
	5	1.07
	4	0.70
	3	0.52
	2	0.30
	1	0.20
	0	0.09
244 (800)	10	3.90
	8	2.82
	6	1.91
	5	1.52
	4	0.92
	3	0.68
	2	0.37
	1	0.24
	0	0.11

Slope Length: m (ft)	Slope (%)	LS Factor
488 (1,600)	10	5.52
	8	3.99
	6	2.70
	5	2.15
	4	1.21
	3	0.90
	2	0.46
	1	0.30
	0	0.12
975 (3,200)	10	7.81
	8	5.64
	6	3.81
	5	3.03
	4	1.60
	3	1.19
	2	0.57
	1	0.36
	0	0.14

Equation for Calculation of LS (if Not Using Table 30)

$LS = [0.065 + 0.0456 \text{ (slope)} + 0.006541 \text{ (slope)}2](\text{slope length} \div \text{constant})^{NN}$

Where:

slope = slope steepness in %

slope length = length of slope in m (ft)

constant = 22.1 metric (72.5 Imperial)

NN = see Table 31 below.

Table 31: NN Values

S	< 1	1 ≤ Slope < 3	3 ≤ Slope < 5	≥ 5
NN	0.2	0.3	0.4	0.5

Table 32: Crop Type Factor

Crop Type	Factor
Grain corn	0.40
Silage corn, beans & canola	0.50
Cereals (spring & winter)	0.35
Seasonal horticultural crops	0.50
Fruit trees	0.10
Hay and pasture	0.02

(Contd.)

Table 33: Tillage Method Factor

Tillage Method	Factor
Fall plow	1.0
Spring plow	0.90
Mulch tillage	0.60
Ridge tillage	0.35
Zone tillage	0.25
No-till	0.25

Table 34: P Factor Data

Support Practice	P Factor
Up & down slope	1.0
Cross slope	0.75
Contour farming	0.50
Strip cropping, cross slope	0.37
Strip cropping, contour	0.25

Table 35: Soil Loss Tolerance Rates

Soil Erosion Class	Potential Soil Loss tonnes/hectare/year (tons/acre/year)
Very low (tolerable)	< 6.7 (3)
Low	6.7 (3) –11.2 (5)
Moderate	11.2 (5) –22.4 (10)
High	22.4 (10) –33.6 (15)
Severe	> 33.6 (15)

Table 36: Management Strategies to Reduce Soil Losses

Factor	Management Strategies	Example
R	The R Factor for a field cannot be altered.	–
K	The K Factor for a field cannot be altered.	–
LS	Terraces may be constructed to reduce the slope length resulting in lower soil losses.	Terracing requires additional investment and will cause some inconvenience in farming. Investigate other soil conservation practices first.
C	The selection of crop types and tillage methods that result in the lowest possible C factor will result in less soil erosion.	Consider cropping systems that will provide maximum protection for the soil. Use minimum tillage systems where possible.

Factor	Management Strategies	Example
P	The selection of a support practice that has the lowest possible factor associated with it will result in lower soil losses.	Use support practices such as cross-slope farming that will cause deposition of sediment to occur close to the source.

Example: Calculation of Soil Erosion Using USLE

$A = R \times K \times LS \times C \times P$

Rainfall and Runoff Factor (R)

The sample field is in Middlesex County. Therefore the *R* Factor is obtained in Table 28 from the London weather station.

R Factor = 100

Soil Erodibility Factor (K)

The sample field consists of fine sandy loam soil with an average organic matter content. The K Factor is obtained from Table 29.

K Factor = 0.40

Slope Length-Gradient Factor (LS)

The sample field is 244 m (800 ft) long with a 6% slope. The LS factor can be obtained directly from Table 30 or may be calculated using the equation. The NN value from Table 31 to be used in the equation is 0.5.

LS Factor = 1.91

Crop/Vegetation and Management Factor (C)

The sample field was plowed in the spring and grain corn was planted. The C Factor is obtained from the crop type factor (Table 32) and the tillage method factor (Table 33).

Crop Type Factor for grain corn = 0.4
Tillage Method Factor for spring plow = 0.9
C Factor = 0.4 × 0.9 = 0.36

Support Practice Factor (P)

Cross-slope farming is used on this sample field. The P Factor was obtained from **Table 34**.

P Factor = 0.75

Therefore,

$A = RK \times LS \times C \times P$
$= 100 \times 0.40 \times 1.91 \times 0.36 \times 0.75$
$= 20.63$ tonnes/hectare/year (9.28 tonnes/acre/year)

Referring to Table 35 in this Factsheet, you will see that this soil loss rate of 20.63 tonnes/hectare/year (9.28 tons/acre/year) is in the moderate range and considerably higher than the "tolerable loss level" of 6.7 tonnes/hectare/year (3 tons/acre/year). To reduce the soil losses for this sample field below 6.7 tonnes/hectare/year (3 tons/acre/year) we will make the following changes to the above example.

Change tillage method from "spring plow (0.9)" to "no-till (0.25)"

Therefore, C Factor (revised) = $0.4 \times 0.25 = 0.10$

The adjusted annual soil loss value is

$$A = R \times K \times LS \times CP$$

$$= 100 \times 0.40 \times 1.91 \times 0.10 \times 0.75$$

= 5.73 tonnes/hectare/year (2.58 tons/acre/year)

Thus by changing the tillage practice, the average annual predicted soil loss for this field is below the "tolerable soil loss" of 6.7 tonnes/hectare/year (3 tonnes/acre/year).

Wind Erosion

Wind erosion is a major problem in arid and semi arid regions of the country which are characterized by mostly dry lands. It is primarily responsible for creation of deserts and sand dunes in the world. Sandy soils are more prone to this type of erosion as the drift of finer soil particles is easier. Along with soil, the organic matter and nutrients can also be removed. Erosion starts with the minimum wind velocity of 15 kilometers per hour and it depends on the velocity of wind and the size of the soil particles.

Estimating Soil Losses by Wind

In order to predict the amount of wind erosion that is likely to occur under a given set of field conditions, attempts have been made to develop a numerical equation combining the factors which influence wind erosion. However, the evaluation of the constants in the equation for wind erosion is comparatively difficult than the universal soil loss equation.

The equation is of the form-

$$E = I\,R\,K\,F\,C\,W\,D\,B$$

Where,

E = Soil loss by wind erosion

I = Soil cloddiness factor representing the ratio of non-erodible soil aggregates

R = Surface cover factor

K = Surface roughness factor

F = Soil textural class factor

C = Factor representing local wind conditions

W = Field width factor

D = Wind direction factor

B = Wind barrier factor

Coshocton Wheel

The Coshocton wheel silt sampler (developed at an experimental station at a place known as Coshocton in USA) can be used for collecting water samples from the runoff passing through a multi-slot divisor or directly through a measuring device like the H-flume (Fig. 7).

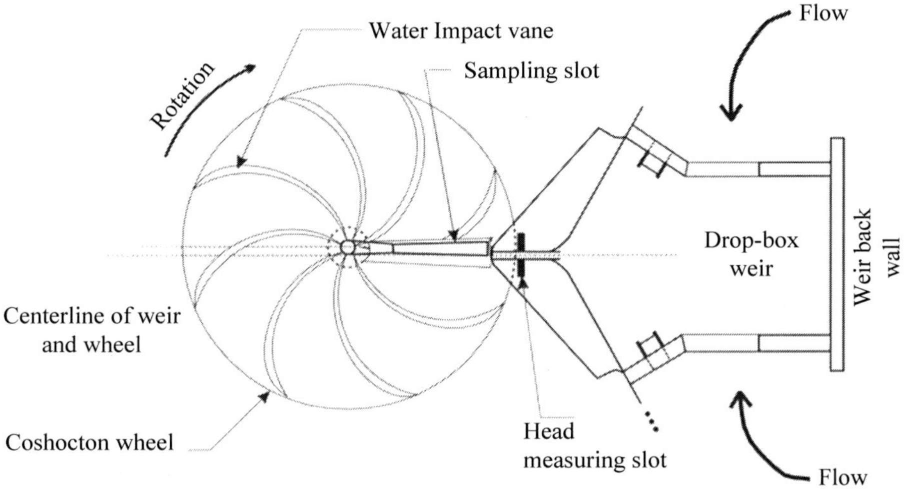

Fig. 7. Schematic diagram of coshocton wheel

The device consists of a wheel made of thin circular plate to which eight vanes are attached. A sampling head with a narrow opening called slot is mounted radially on this wheel. The slot is kept raised sloping down towards the outside of the wheel.

A cylindrical pan below the water wheel receives the sampled portion of the runoff. An outlet pipe from the pan leads to a collecting tank where the sample runoff gets collected. When the discharge from the measuring flume falls on the wheel its impact rotates the wheel. With each revolution of the wheel, the slot cuts across the jet from the flume and extracts a small portion of the flow. The extracted portion is collected in the storage tank through the conduit. An expression for the relation between rate of flow over the sampler and the flow collected by the sampler is derived as follows:

A jet of water with flow Q, width x, thickness y and velocity v (Fig. 8) is allowed to fall over a slot with length greater than y and with width w. The

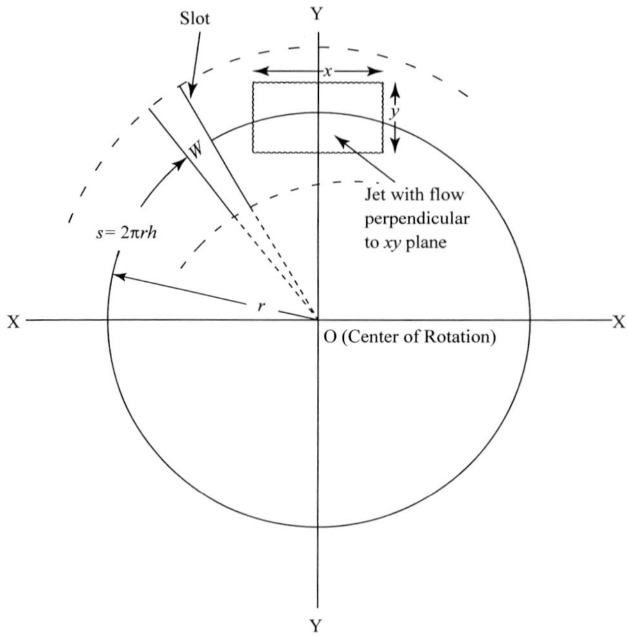

Fig. 8. Mechanics of flow in a coshocton wheel

discharge Q_1 into the slot is:

$Q_1 = y.\ w.\ v$

When the slot is rotated about Q_1 with a speed S, the time taken by the slot to cross the jet completely is given by:

$t = x/S$

where,

$S = 2\pi$ rn,

n = rotational speed of the wheel containing the slot,

r = radius of the wheel.

The volume of water for one pass through the jet

$= Q_1.\ t = wyvx/2\pi rn$

The volume (q) extracted in unit time is the amount caught in n passes, *i.e.*,

$Q = wyvx/2\pi rn = wQ/2\pi r$ ($\because Q = yvx$)

The ratio of the mean flow from the slot to the total flow of the jet is:

$R = q/Q = (wQ/2\pi r)/Q = w/2\pi r$

The ratio R is called the sampler ratio. It is the ratio of the slot width to circumference at any radius. It may be noted that R is independent of the speed of rotation of the wheel.

Therefore, the Coshocton type runoff sampler can also be used to compute the total runoff from the sampler ratio provided q can be recorded. Because of the difficulty in recording q, these samplers are used only for sediment sampling and Q is measured using some rate measuring flumes.

Since the runoff and the average silt concentration are known, the total soil loss during the particular storm can be computed. The H-Flume and the Coshocton sampler in combination are used for runoff and sediment estimation from small agricultural watersheds.

Stage level recorder

Water-stage recorders consist of a group of instruments that produce a record of water surface elevation with respect to time. The output can be analog (providing a graphical result) or digital (punched paper tape or stored or transmitted values). Important advantages of recorders over non-recording staff gauges are:

1. In streams having daily fluctuations, continuous records provide the most accurate means of determining the daily average gauge height.
2. Maximum and minimum stage are recorded, and the time they occurred can be noted.
3. Records can be obtained at stations where observers are not always available.

(a) Analog–Graphical Recorders

In general, analog or graphical recorders consist of two main elements: a clock mechanism actuated by a spring, weight, or electric motor and a gauge height element actuated by a float, cable or tape, and counterweight. Four basic types of recorders use these elements. Fig. 9 shows a horizontal drum recorder, in which the clock positions the pen along the drum axis, and the gauge height element rotates the drum. This recorder is also available with a vertical drum. Another type of recorder also has a vertical drum, but the time and height elements have been reversed so that the clock mechanism rotates the drum.

Fig. 9. Horizontal drum water-stage recorder. The time element records parallel to the axis of the drum. (*Courtesy*: Leupoild and Stevens Instruments, Inc.).

These types of recorders usually operate using 8-day, spring-driven clocks. Electrical drives could also be used if a reliable *Source* were readily available. The stylus, usually either a capillary pen or a pencil of proper hardness, must be capable of operation for the full 8 days without attention. To accommodate various water-stage differentials, ratios of water-stage change to recorder-chart change are available from 1:1 to 10:1 and should be specified at the time the recorder is ordered. The standard width of recorder paper is 10 inches (in), and all recorders come equipped with metal covers.

The fourth type of graphic recorder is shown on Fig. 10 and 11. The time element, consisting of a compensated, balanced, weight-driven clock, drives two parallel rolls, one of which holds the supply paper. The paper unrolls from the supply roll at a uniform rate and with constant tension and is taken up on the receiving roll. Speed of travel may be adjusted from 0.3 to 9.6 in per day on any standard instrument, and other chart speeds are available on special order. The normal chart length is 75 feet (ft).

Fig. 10. Continuous recording water-stage recorder with cover raised. The time element rotates the rolls, and the height element records parallel to the axis of the rolls.
(*Courtesy*: Leupoild and Stevens Instruments, Inc.).

The float activates a pen stylus which moves parallel to the axis of the rolls so that 1 in of travel represents a change in water stage of 1 ft. The stylus is designed so that it can be accurately set for gauge height. The ratio of water surface change to stylus travel can also be adjusted to accommodate small to large ranges of depth. The range of the recorder is limited only by the length of the float cable because the stylus reverses direction at the point of maximum deflection. Capacity of the ink reservoir is sufficient for the recorder to operate for 60 days or longer.

Fig. 11. Typical installation of a continuous water-stage recorder in a wooden shelter.

(b) Digital Recorders

Digital recorders used in water stage measurements usually include two types: punched-paper tape (Fig. 12) and analog-to-digital data loggers. Both types are electrically operated (usually by batteries) and record numbers either on the paper tape or in memory at selected time intervals.

Fig. 12: Digital recorder.

Water stage is transmitted to the punched-paper tape recorder, usually through shaft rotation on a float and pulley arrangement. Shaft rotation is converted by the recorder into coded punch-tape records (Fig. 13). The code consists of four groups of four punches each. In each group, the first punch represents "1"; the second, "2"; the third, "4"; and the fourth, "8." Thus, a combination of 1, 2, or 3 appropriate punches in a given group represents digits from 1 to 9. A blank (no punch) represents zero. Together, the four groups of punches represent all numbers from 0 to 9999.

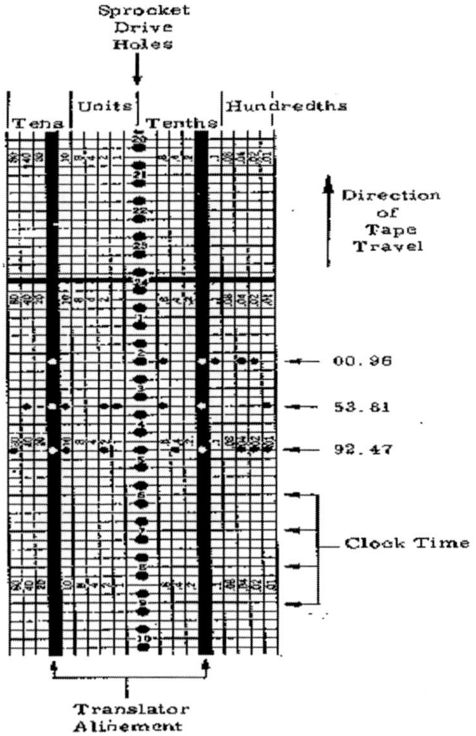

Fig. 13. Digital recorder punched tape. (*Courtesy:* U.S. Geological Survey).

Coding is done by either one or two identical disks that have raised ridges on the faces. Fig. 14 shows a digital recorder with only one disk, which produces 2-digit numbers ranging from 0 to 99. Two disks would be used if 3 or 4 digit numbers are required. The right-hand disk is connected directly to the input shaft, and the left-hand disk is driven from the first disk by a 100 to 1 worm-gear reducer. One-hundred revolutions of the input shaft and the first disk cause one revolution of the second disk. A paper tape is moved upward through the punch assembly in the center of the instrument. The punch block contains a row of 18 pins or punches C16 for information and 2 for feeding the tape. At the selected intervals of time, the punch assembly is pivoted on a shaft at the bottom, so the punch, paper, and pins are moved toward the disks. Those pins that strike raised ridges on the disks are forced through the paper, punching neat round holes. The pins that do not strike ridges do not punch holes. This record reflects the position of the disks, which is proportional to the water stage, all as a function of time.

The mechanically punched tape is still widely used at this time (1996) and is considered very practical for field use where temperature, moisture, and power conditions are widely variable. Electronic translators can convert punched tape records into suitable input for digital computers.

The most recent advances have been in the area of data loggers. This group of electronic instruments has evolved quickly over the last 10 years. Small, battery–operated, fully programmable units offer many features in addition to data recording. Some type of transducer is required to sense the water stage with this type of recorder. Options range from a pressure transducer sensing water level in a stilling well to a pressure transducer on a bubbler system, to an angular transducer sensing shaft rotation on a float-driven system. In all cases, an analog (voltage or current) output is sensed, digitally recorded, and stored by the data logger. This type of system is perhaps best suited to the transmission of data via satellite or microwave, providing a central control location with current water stage information. Remote sites are very well suited to being powered by batteries which are charged through small solar cells. However, vandalism of the solar panels can be a problem because they must be exposed. Concealing solar panels in some way (such as in the top of a tree) can help.

Necessity of soil and moisture conservation measures in dry lands

The problem of soil erosion is assuming a serious proportion and it is necessary in dry lands as the pressure on land increases. As a great scope exists to stabilize and improve crop yields under dry land conditions, it is obvious to observe that there is a great need for soil and water conservation measures.

A careful study on Planning for Agricultural Development "by Appa Rao (1990) revealed that among several constraints, two key resources-land and water play a very dominant role in agricultural development of various agro climatic zones in India. The nature of damage caused by various problems connected with land and water varies from zone to zone and in most of the zones the predominant constraint is soil erosion and its associated problems Various strategies recommended to overcome such problems emphasize the need for different soil and water conservation measures in different agro climatic regions of India (Table 37).

Table 37: Soil and water conservation measures in different agro climatic regions of India

S. No.	Agro climatic zone	Land and Water related problems	Strategies recommended
1.	Western Himalayas	Landslides, pollution, receding glaciers and water scarcity	A more rational land use planning is required
2.	Eastern Himalayas	Shifting cultivation, Soil degradation	Afforestation and terracing in hills, quality seed production
3.	Lower Gangetic Plains	Adequate storage of ground water with high water table, water logging and marshy lands	Improvement in rice farming, horticulture, pisciculture, poultry, livestock, forage production and seed supply
4.	Middle Gangetic Plains	Flood prone areas, under utilization of ground water	Introduction of modern market oriented agriculture, water harvesting structures and alternative farming systems
5.	Upper Gangetic Plains	Saline/alkaline soils	Developing multiple mixed cropping patterns and water management
6.	Trans Gangetic Plains	Water logging, salinity, alkalinity, soil erosion and falling water table	Water management and reclamation
7.	Eastern Plateau & Hills	Rain water runoff and decrease in ground water potential	Water management, maximise use of rain water, integrated watershed development approach
8.	Central Plateau & Hills	Dry land agriculture, ravines and soil erosion	Watershed development, water conservation through water saving devices
9.	Western Plateau & Hills	Deficient in water resources	Popularizing water saving devices
10.	Southern Plateau & Hills	Dry land farming	Watershed approach, restoration of tank irrigation
11.	East Coast plains & Hills	Water logging and alkalinity.	Improving tank irrigation
12.	Western Plains & ghats	Salinity ingress	Protection of land from salinity ingress and provision of Drainage

S. No.	Agro climatic zone	Land and Water related problems	Strategies recommended
13.	Gujarat Plain & Hills	Under exploitation of ground water and rain water run off	Canal and ground water management, rain water harvesting, agro-forestry development, wasteland development
14.	Western dry region	Low rainfall region, drought	Moisture conservation structures
15.	Islands	Rainwater run off	Water management and fisheries

Source: Anon., 2019

Soil and Moisture Conservation Practices in Dry Lands

In India, about 87 million hectares of cultivated lands receive a mean annual rainfall in the range of 500 to 750 mm which makes the crop production in these areas dependent entirely on the onset, quantity, frequency and distribution of rains. The monsoon rains are scanty, highly erratic and often ill distributed. It is thus necessary to conserve the moisture in dry lands.

Factors influencing soil moisture storage

1. Slope of the land which decides the run off and its quantity
2. Soil physical properties especially soil texture and structure decides the storage of moisture in soil. For example in a coarse textured soil the moisture conservation will be less and percolation will be more.
3. Chemical properties of soil like soil pH and cation exchange capacity
4. Depth of soil which decides the quantity of soil moisture storage
5. Climatic conditions like rainfall, its quantity and distribution, temperature, humidity *etc.*
6. Presence of hard pan in plough layer or beneath the layer will reduce the storage capacity of soil.

Features of Soil Moisture Management

Land management practices and crop management activities can control soil erosion and also improve the water conservation in agricultural lands. Land management practices are often referred as mechanical measures or engineering structures and different crop management activities aiming at soil and water conservation are called as agronomic or biological measures. The different features of soil and moisture conservation are as follows.

1. Control of erosion by suitable mechanical structures.
2. Storage and conservation of rainwater by tillage and suitable *in-situ* measures
3. Efficient utilization of conserved moisture by appropriate soil and crop management practices.
4. Collection and storage of excess runoff in farm ponds for supplemental or protective irrigation.
5. Alternate crop planning or management practice depending upon the rainfall

Techniques of soil and water conservation

Techniques: refers to conserving of rain at a place where it is received. Agronomic practices have to be considered in co-ordination with others (mechanical/agrostological) for their effectiveness. They are effective on gentle slope (up to 9%).

The objective of soil moisture management is to control soil, water and wind erosion at various levels. It is aiming on enhancing the moisture availability period and crop growing season. Soil moisture can be improved either by reducing the losses or enhancing the water uptake and use efficiency. The conserved water will be efficiently used by the selection of compatible crops and cropping systems, less water intensive crops, higher harvest index crop, higher root/shoot ratio (Shankar and Premalatha, 2005).

Soil and Moisture Conservation Practices (*In situ* and *ex situ*)

Three aspects of soil and moisture conservation need special emphasis:

i. To prevent soil erosion, as soil is the medium of storage
ii. To increase *in-situ* moisture conservation
iii. To facilitate smooth disposal of surplus water to avoid the ill effects of water stagnation

In-situ rain water management practices aims on conserving the available water inside the cultivated field by reducing runoff and improving infiltration. Selection of suitable conservation measures depends on soil type/parent material, rainfall pattern, slope, vegetation, crops and intensity of erosion. There are various types of *in-situ* moisture conservation measures with which the soil moisture can be conserved and the crops can be raised successfully under dry land conditions.

The important practices for *in situ* moisture conservation are

1. Cultural/Agronomical methods

i. Addition of organic manure and fertilizers

Addition of farmyard manure, organic manure, green and green leaf manures help to increase the water holding capacity of the soil and to avoid formation of crust in redsoils. Fertilizers improve vegetative growth (canopy) which aids in erosion control. Improvement in soil structure improves rate of infiltration leading to reduced runoff.

Dry lands are hungrier than thirsty. Due to poor returns, the depleted nutrients are not often replenished. Added to this, specific nutrients may be limiting in a particular soil. Unless this limiting nutrient is identified and corrective measures adopted, application of other inputs may not lead to any increase in yield.

Plant nutrients can be supplied with different sources *viz*., organic manures, crop residues, biofertilizers and chemical fertilizers. For better utilization of resources and to produce crops with less expenditure, integrated nutrient management is the best approach in which all the possible source of nutrients are applied based on economic consideration and the crop is supplemented with chemical fertilizers to balance the required nutrients.

Under low fertility conditions, fertilizer application brings about increases in root growth and proliferation. This helps to utilize soil moisture from larger soil volume and thus imparts tolerance to drought effectively. Fertilized crops mature early and produce more per unit of water consumed. Fertilizer is a costly input; hence improving the fertilizer use efficiency is extremely important for resource poor farmers. Some major points to be kept in view to improve the fertilizer use efficiency are:

- Fertilizer should be placed in bands, preferably by seed-cum-fertilizer drill.
- Nitrogenous fertilizer should be applied in splits.
- Fertilizer composition should be balanced to suit the crop requirement.
- To reduce the expenditure on fertilizers and to improve their uptake, organics like FYM, compost, crop residues, green manures *etc*., should be applied liberally.
- Bio fertilizers, wherever recommended, should be invariably made use.

The outcome of the research findings of dryland Agriculture project, Bangalore centre indicates:

- For sustaining higher level of productivity and maintaining soil health, an integrated nutrient management (INM) system involving both organics and inorganics is essential (Table 38).

- Ill effects, if any of unbalanced fertilization can be overcome by continued application of organic manures like FYM.

Table 38: Influence of long term use of FYM and Maize residue (MR) on productivity and economics of finger millet under finger millet mono-cropping

Treatments	Yield(kg ha^{-1})			Cost of cultivation (Rs ha^{-1})	Net returns (Rs ha^{-1})	B:C ratio	RWUE (kg ha-mm^{-1})	SYI
	Grain	Mean grain yield (41 years)	straw					
Control	93	645	145	12480	–9008	0.24	0.34	–0.09
FYM (10 t/ha)	1545	2222	1624	20132	36379	2.37	5.59	0.44
MR (5 t/ha)	362	965	422	15480	–2177	0.86	1.31	0.10

Source: Anon., 2019

RWUE: Rain Water Use Efficiency

SYI: Sustainable Yield Index

The finger millet RWUE, grain and straw yield was found to be higher (5.59 kg ha-mm^{-1}, 1545 kg ha^{-1} and 1624 kg ha^{-1} respectively) with application of FYM @ 10 t ha^{-1} compared to application of maize residue 5 t ha^{-1} (1.31kg ha-mm^{-1}, 362 and 422 kg ha^{-1} respectively) and control (0.34 kg ha-mm^{-1}, 93 and 145 kg ha$^{-1,}$ respectively) under rainfed conditions in *kharif* season.

Technology developed at AICRPDA, Bangalore centre clearly indicated the usefulness of *in-situ* incorporation of horse gram in the early *kharif* season with 20–22 tonnes of biomass per hectare and 30 quintals of finger millet per hectare after horse gram with 50 per cent saving in recommended N. Thus, this technology not only improves the organic matter status of dryland but also improves the physical and chemical properties of soil apart from saving of 50% N (Anon., 2008). Further *ex–situ* incorporation of glyricidia grown on bunds and boundaries (300–325 plants ha^{-1}) produced green biomass to supply 25–30 kg N ha^{-1} besides atmospheric nitrogen (30–50 kg N ha^{-1}) (Anon., 2007).

In a long range study, continuous use of green leaf manure to supply 50% N + 50% N and 100% PK (fertilizers) in finger millet/Soybean/Maize rotation system resulted in higher finger millet yield (3004 kg ha^{-1}), maize (2790 kg ha^{-1}) grain yield as compared to application of recommended NPK (2572 and 1150 kg ha^{-1}, respectively) (Anon., 2009). Application of 12.5 kg $ZnSO_4$ per hectare + 10 kg Borax per hectare along with recommended NPK increased the seed yield of finger millet by 15–21% (3500 kg ha^{-1}) (Das, 1985).

in situ Green manure incorporation

Introduction of rotovator takes care of operation like, uprooting of plants, chopping and incorporation in the soil simultaneously and maintains a uniform seed bed. It facilitates decomposition of incorporated plants like horse gram, cowpea, sunhemp *etc*. Hence, energy costs are reduced to 50–60% as compared to other implements used for this operation. It can cover one hectare in 90 minutes by using 35 HP tractor and having yield advantage of 32.6% with additional income of ₹ 4,112/- per hectare.

ii. Summer ploughing/Off season tillage: Plough furrows can hold water in the depressions and thereby increase the infiltration. When done across the slope, the plough furrows check runoff, reduce the velocity of runoff water and improve storage. Summer tillage is a traditional practice helps in the storage of pre-sowing rainfall. When ploughing is done along contour it is termed as contour ploughing and is more helpful for *in situ* moisture conservation. Summer ploughing also helps in control of perennial weeds, pest control and enables early sowing with onset of rains.

Principles of summer ploughing in dry lands

1. To reduce or prevent soil erosion
2. To improve the soil fertility and moisture absorbing capacity of soil
3. To reduce or arrest weed growth
4. To control insects

Advantages of summer ploughing in dry lands

1. It reduces cracking in rain fed vertisols, thereby preventing moisture loss through evaporation from deeper layers of soil
2. It facilitates early decomposition of crop residues in soil
3. It controls the weed germination and their growth and thereby reducing the nutrient and moisture loss through weeds
4. It prevents the flowering and seed formation in weeds and their further spread under cropped conditions
5. It arrests or reduces soil erosion due to subsequent summer showers
6. It improves the moisture absorbing capacity of the soil
7. It brings out and destroys certain soil borne insects
8. It increases the yield potential of pre monsoon sown crops

Off season tillage is an important component of rainfed farming technology especially useful in red soils and other related soils if taken up soon after the harvest of previous crop or just after receiving the early showers. Essentially it is meant to open up the soil for better water intake with the summer rains. The operation has dramatic effects especially in sub-normal years. The physical manipulation of surface soil should not destroy physical and chemical properties of soil. Tilth is the physical condition of soil in its relation to plant growth. In drylands, tillage is required to (i) improve moisture conservation:

(ii) Break hard pan and help root penetration: (iii) Control weeds: (iv) Provide better aeration and (v) Good seed bed. The soil aggregates stability depends upon the extent of disturbance by mechanical force during the course of tillage (**Table 39**).

Table 39: Effect of mould board ploughing on the yield of finger millet (kg ha^{-1})

System	Average of two farmers in each year	
	1981	1982
Local practice (2–3 wooden ploughings)	1220	1040
One additional MB ploughing in July	1670	1570

Source: Anon, 1982

Ploughing in July with mould board plough would enhance the productivity of finger millet substantially over 2–3 wooden ploughings.

One survey conducted by the team of Operational Research Scientists in Bangalore showed that more than 50% of the farmers do not own even a bullock pair. Custom hire services or societies owing required machinery, can work. Further, what is required is affordable power equipment, which small farmers can buy and maintain individually.

Soil crusting

Bare soil is susceptible to damage from rain drop impact. High energy rainfall impacting on bare soil, breaks down aggregates and compacts a thin surface layer, leading to seal or crust formation and hard setting. Crusting is sometimes used to describe surface seals that have dried, but more often is used interchangeably with sealing. Surface sealing and crusting differs from hard setting in that rainfall impact is required for a seal or crust to form, whereas hard setting occurs without any externally applied load. Hydraulic conductivity of seals can be as much as 2000 times less than unsealed soil (Melntyre, 1958). Hydraulic conductivity of seals decreases exponentially with cumulative rainfall energy and time taken to reach steady state value as influenced by surface conditions. If a crust forms on drying, subsequent infiltration gets depressed and may persist until the crust is disturbed by tillage (Connally, 1998).

Crust management in dry lands (Gajanana, 2005)

- Breaking the crust soon after formation with a crust breaker or by periodic deep ploughing.
- Improving organic matter content of the soil, by applying FYM, Compost, crop residue and green manure.
- Applying gypsum in sodic black soils and phosphorous fertilizer in red soils in seed rows.

iii. Tillage

Deep tillage for chiselling and fall ploughing to break the top hard layer (top crust in red soils) so as to make the soil responsive for the monsoon rains help in increasing infiltration and moisture retention and thus increase crop yields. The beneficial effects of deep tillage are found to continue for two years. Hence, deep tillage is recommended once in three years while fall/summer ploughing is recommended every year/season in lighter soils. About 20–25% yield increase has been recorded in dryland crops under deep tillage compared to shallow tillage at Ananthapur. Tillage and sowing on contour, and opening furrows in between crop rows help in inter-row harvesting of rain water and increase the time of concentration and moisture retention. Tillage also helps in uniform distribution of soil moisture as well as its longer period of availability. Experiments in red soils in Kabbanala watershed in Karnataka have shown that contour cultivation not only increased crop yields by about 25–28% in sesame, finger millet and groundnut but also increased the fertilizer use efficiency by about 39–58% (Table 40).

Table 40: Depth of tillage on crop yield in Alfisols

Crop	Crop yield (kg ha^{-1})			
	Wooden plough (5–10 cm)	Light hoe (10–15 cm)	MB plough (10–20cm)	Chisel plough (20–30 cm)
Groundnut	1176	1045	1202	1036
Finger millet	2718	1507	2807	2943
Maize	3555	4057	4223	4428
Redgram	1231	–	1563	1892
Sunflower	1094	1338	1002	1483

Finger millet, maize and redgram are beneficial with deep tillage whereas sunflower and groundnut are benefited by medium tillage.

Table 41: Percentage soil moisture after 81 mm rain during May in red soils of Bangalore

Depth of soil (cm)	Moisture percentage on oven dry basis	
	Ploughed	Unploughed area
0–15	10.74	3.6
15–30	13.22	7.13
30–60	13.77	5.89
60–90	13.44	Dry

Source: Shankar and Premalatha, 2005

The concept of minimum and reduced tillage was more popular in western countries and highly mechanized plot. The concept of minimum or zero tillage may not work under Indian conditions, except on Vertisols of certain areas and to certain extent under irrigation. Minimum tillage may not permit storage of as much of rain water as possible in soil profile for subsequent use by the crop(Table 41).

iv. Contour farming: Ploughing and all other operations along the contour (across the slope) and sowing reduce soil erosion and reduce runoff. Contour farming or tillage is the practice of carrying the field operations such as ploughing and cultivating the land across the slope. As such the operations are done at right angle to the slope of the land, each ridge will intercept the flowing water and contoured rows hold the water and thus the run off water and soil erosion are prevented. In steeper slopes when contour farming alone is done, there may be soil erosion as there is a possibility of breaching in rows. It will lead to cumulative damage and poor infiltration of water. On steeper slopes, contour farming should be supplemented by strip cropping, bunding or terracing (Fig. 14).

Fig.14. Contour farming cum Vegetative bund

v. Cover crops: A cover crop may be defined as a crop which serves as a solid ground cover, planted mainly to protect the soil between tree crops or semi permanent crops or between the seasons for regular crops, Generally they are grown for reducing soil erosion, improving soil productivity and adding organic matter to the soil. The best cover crop should be shade tolerant, easily propagated and require least management.

Erosion is reduced if the land surface is fully covered with foliage. Eg., Black gram, green gram, Groundnut, Fodder grasses like *Cenchrus ciliaris*, Both contour cropping and cover cropping can be practiced when the slope is less than 2%.

Cover crops are those crops which have fast initial growth and cover the land immediately which prevent the soil from beating action from rain and control the runoff and soil erosion. Vegetation is crucial in preventing erosion of productive soil and conserving rain water for sustaining life. Soil and water needs to be preserved for crop production. Crops like green gram, black gram, horse gram and cowpea are some of the covers crops.

Cover crops in Black soils

Growing of cover crops helps in harvesting rainwater in-situ effectively and reduce the splash erosion. Apart from this, the farmers will get an additional income ranging from 43% to 300% over *kharif* fallow (Table 42).

Table 42: Yield of crops as influenced by cover crops in Black soils

Production system	Yield (kg ha^{-1})		Rabi sorghum equivalent yield (kg ha^{-1})	% increase over farmers practice
	Kharif crops	*Rabi* crops		
Cucumber-*rabi* sorghum	6250	1750	2583	250
Greengram-sunflower	400	750	2717	50
Cucumber-sunflower	6250	788	2567	43
Ridgegourd-*rabi* sorghum	375	2000	2050	300
Ridgegourd-sunflower	1250	1650	3797	200
Bittergourd-*rabi* sorghum	1000	1800	2133	260
Fallow-*rabi* sorghum (farmers practice)	–	500	500	–

Source: Ramachandrappa *et al.* (2014)

Advantages of cover cropping

1. It reduces run off and conserves moisture
2. It prevents soil erosion
3. Cover cropping increases water infiltration and water absorbing capacity of soil
4. Cover crops add organic matter to the soil and improves soil tilth with deep root system
5. It protects newly constructed terraces and other mechanical structures for the control of soil erosion
6. A cover crop provides fodder during winter
7. It adds nitrogen to the soil

Table 43: Crops suited for cover cropping

S. No.	Common name	Botanical Name	Remarks
1.	Calapo	*Calopogonium mucunoides*	Perennial
2.	Sword Bean	*Canavalia ensiformis*	Annual
3.	Pigeon Pea	*Cajanus cajan*	Annual
4.	Centrocema	*Centrocema pubescens*	Perennial
5.	Desmodium	*Desmodium intortum*	Perennial
6.	Creeping Indigo	*Indigofera endecaphylla*	Perennial
7.	Lablab Bean	*Lablab purpureus*	Annual
8.	Velvet Bean	*Mucuna capitata*	Annual
9.	Stylo	*Stylosanthes guinesnsis*	Perennial
10.	Cowpea	*Vigna sinensis*	Annual

vi. Cropping systems: Monocropping of erosion permitting crops accelerate soil and water loss year after year. Intercropping erosion permitting and erosion resistant crops or their rotation have been found effective for soil and water conservation. As the legumes (cowpea, greengram, horsegram, blackgram) are effective for soil conservation due to their smothering effect, they should be sown in time to develop adequate canopy by the time of peak rate of runoff.

Strip cropping

Strip cropping can be defined as a method of growing soil conserving and soil depleting crops in alternate strips running perpendicular to the slope of the land or to the direction of the prevailing winds for the purpose of reducing wind and water erosion. It is an effective method of erosion control under mild rainfall and on gentle slopes of 7–12%. In heavy rainfall regions, it should be supported by structural measures. There are generally four types of strip cropping which are described as in Table 44.

The degree and length of slope, soil permeability, susceptibility to erosion, quantity and intensity of rainfall and crop rotation are a few factors that decide the width of the strips. In dry farming areas, it is desirable to have strips of width 12 to 320 feet for the purpose of controlling wind erosion. Having the strips too far will lead to soil and water losses and therefore depending upon the soil type, cropping, tillage system and wind velocities, the width of the strips can be decided.

Precautions in strip cropping

1. Proper crop rotation should be maintained in strip cropping
2. Strips should have uniform ridges and furrows
3 The selected crop should yield an additional benefit like forage or food for easy acceptance by the farmers

Table 44: Types of strip cropping

S. No.	Types	Uses (to control)	Areas	Method of cropping
1.	Contour strip cropping	Water and wind erosion	Slope lands	Arranging crops in strips/bands on contour at right angle to the slope
2.	Field strip cropping	Water erosion	Irregular topography	Cultivate crops in strips of uniform width across the general slope
3.	Wind strip cropping	Wind erosion	Leveled	Growing crops in strips of windy areas, uniform width with straight and at right angle to the prevailing wind direction
4.	Buffer cropping	Water and wind erosion	Steep eroded slopes	Laying out crops in strips of grass or a legume in between strips of regular crops

(vii) Mulching: Mulching reduces soil loss considerably by protecting the soil from direct impact of rain drop and reducing to keep sediment carried out runoff. A minimum plant residue ground cover of 30% is necessary to keep runoff and soil loss within acceptable limits. As most of the plant remains are used as cattle fodder, availability of organic residues for mulching is the major limitation under conditions of scarce fodder.

An *in-situ* moisture conservation practices of retaining crop residues on the soil surface. After harvest of the *kharif/rabi* crops, all the residues of sunflower, pigeon pea straws with leaf litter, stubble of sorghum, straw of chickpea *etc*., are left on the soil surface. These residues act as a barrier for the flow of runoff, help in more infiltration and increased soil moisture status by providing more opportunity time for water to infiltrate the soil.

In the summers, the aim of mulch is to conserve moisture while in winters, it is used to conserve soil heat and prevent frost damage. Different kinds and their effects are described below (*Source*: Takatori *et al.,* 1964)

S. No.	Type of mulching	Material used	Uses
1.	Organic mulching	Saw dust, straw, rice husks, crop residues *etc.*	Moisture conservation Increases nutrient availability Less soil crusting Reduces soil temperature

S. No.	Type of mulching	Material used	Uses
2.	Chemical mulching	Aluminum foils, petroleum products like plastics, polythene sheets	Reduces/increases soil temperature based on type of material used. Involve High cost
3.	Soil mulching	Practice of frequent inter-cultivation	Prevents loss of moisture from deeper layers

Advantages of mulching under dry lands

1. It reduces the loss of moisture through evaporation and thus conserves soil moisture
2. It prevents or reduces the germination and growth of weeds
3. It helps in maintaining the soil fertility
4. It can reduce the root injury and soil borne diseases
5. It improves soil physical properties
6. It controls surface run off and prevents soil erosion

Disadvantages of mulching

1. Effect of the practice depends on the climate, season and soil type
2. Mulching may not be advantageous in all places
3. The problem associated with mulching are occurrence of pest and disease
4. Availability of mulch materials in large quantities will be doubtful (for Eg. 5 tonnes of straw per hectare of land)

Mulching crop residues in dryland crops reduces evaporation, conserve soil moisture and reduce the damage caused by dry spell (Table 45, 46 and 47).

Table 45: Effect of surface mulch on yield (kg ha^{-1}) of sorghum

Treatment	Grain yield (kg ha^{-1})	straw yield (kg ha^{-1})
Control	720	3430
Sorghum stubbles	1020	4490
Redgram stalks	1210	5520
Wheat straw	1040	5340
Dry grass	1220	5750

Source: Patil *et al., 1981*

Frequent intercultivation between crop rows to create a dust mulch is a useful operation that helps in breaking soil crust (especially in red soils). It augments high infiltration and breaks the capillary movement of water to the top layer and minimizes evaporation losses. Research studies at Bellary have indicated the possibility of doubling of water-use efficiency and crop yields by providing dust mulch through repeated harrowing with a bullock-drawn harrow (Table 46).

Table 46: Influence of dust mulch on water use efficiency and grain yield of Pearl millet

Treatment	Moisture used (cm)	WUE (kg ha^{-1} cm^{-1})	Yield (t ha^{-1})
Dust mulch through harrowing	27.8	67.4	1.74
No mulch	22.4	30.7	0.81

Source: Ram Mohan Rao (1981)

Vertical mulches by embedding sorghum or maize stalks in rows at regular intervals across the slope is a useful practice in black soils, especially those with low infiltration rate. Long-term studies of the dryland project at Bijapur showed that sorghum yields under vertical mulching at 5 m intervals were about 25% higher than control. However, this beneficial effect was not observed when such mulches were used at 10 m intervals (Table 47).

Table 47: Influence of vertical mulching on *Rabi* sorghum

Treatments	*Rabi* sorghum yield (kg ha^{-1})
No vertical mulching	1310
Vertical mulching at 5 m interval	1650
Vertical mulching at 10 m interval	1240

Source: Itnal (1981)

- Vertical mulching for 1–3% sloppy lands
- Increase in yield to the tune of 38–40%.

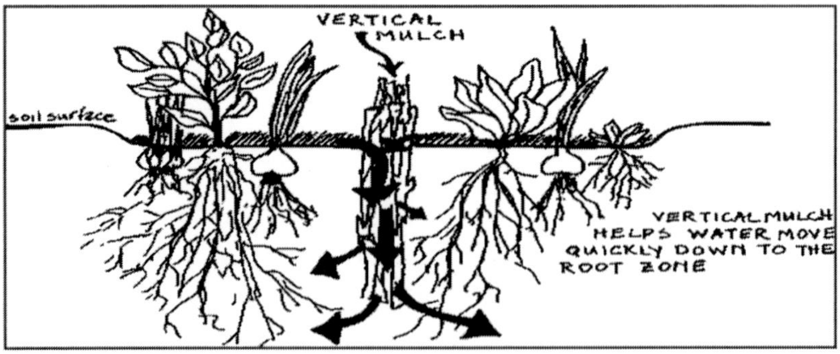

Another important finding found extremely useful, is organic mulching. One experiment conducted at ICRISAT during 1980's is taken here only to highlight the striking effects (Table 48).

Table 48: Effect of groundnut shell mulch on crop yield, runoff and soil loss of *Alfisols* at ICRISAT centre, Patencheru (1981–85)

Rate of Groundnut shell mulch (t ha^{-1})	Crop yield (t ha^{-1}) (intercropping)		Runoff (mm)	Soil loss (t ha^{-1})
	Sorghum	Pigeonpea		
No mulch	2.07	0.84	282	3.92
2.5	2.27	0.97	147	1.95
5.0	2.42	1.11	88	1.21
10.0	2.63	1.34	26	0.35

Source: Pathak, 1984

Several such examples could be quoted from across the agro-ecological zones of the country. Yet the technology is not a practice by the farmers. The reason is 'sufficient mulching material is not available'. But is it a statement made after making honest efforts to collect the weeds removed from cropped fields, harvesting green manure plants like glyricidia, gathering crop residues and leaf litter from trees on the farm bunds and roadsides. By citing these examples we are only trying to point out that if honest efforts are made organic mulch material can be harnessed.

Table 49: ***In situ*** moisture conservation through tied ridges and mulching in chilli varieties

Variety	Dry fruit yield(kg ha^{-1})				RWUE (kg ha–mm^{-1})	Net returns (₹ ha^{-1})	B: C ratio
	2018–19		Mean of 7 years				
	With mulch	Without mulch	With mulch	Without mulch			
Samrudhi	1830	1072	1418	996	6.7	1,19,545	5.5
Chikkaballapur local	729	416	914	675	3.5	50,044	2.9

Source: Anon., 2019

Mulching with crop residues in the furrow recorded higher dry fruit yield (1830 and 729 kg ha^{-1}, respectively), RWUE 6.7 kg ha mm^{-1} compared to growing chilli without mulching (1072 and 416 kg ha^{-1}, 2.9 kg ha mm^{-1} respectively). The overall increase in yield was 41–42% (Table 49).

viii. Use of chemicals: Stability of soil aggregates can be increased by spraying chemical like polyvinyl alcohol @480 kg ha^{-1}. Soils treated with bitumin increase the water stable aggregates and infiltration. Organic matter, FYM, green manure and crop residues also increase the stability.

ix. Improved crop production practices: Improved crop production practices are necessarily to be adopted in lands already treated for soil and moisture conservation. The most important aspects to be considered for higher yields are:

- Early sowing
- Optimum plant population
- Seed hardening
- Dry seeding
- Intercultural operations
- Disease and pest management

Early sowing: Early sowing is advantageous under both *kharif* and *rabi* situations. General advantages of early sowing are: (i) better moisture availability, (ii) good seedling vigor and (iii) longer growing season. More specifically this practice helps in avoiding certain pests like shoot fly and ear head bug in *jowar* and gall fly in rice.

Optimum plant population: With the introduction of hybrids and high yielding varieties selected for higher population, higher plant density is being recommended. Establishing an optimum population is made easier through use of seed-cum-fertilizer drill. It may be desirable to use a higher seed rate and adopt early thinning to obtain the desired crop stand. Within optimum population range, wider row spacing, while maintaining the population, is convenient for intercultural operations and ensure moisture supply for a longer period.

Seed hardening: It refers to the technique of subjecting seeds to soaking in water or appropriate chemical solutions and drying in shade. This results in remobilization of nutrients in the embryo and thickening of protoplasm within the body of the seedling, after germination to impart drought tolerance.

Dry seeding: It is the technique of placing the hardened seeds in a dry soil. The seeds are placed at a depth of 5 to 7.5 cm so they germinate at the onset of monsoon and not with the pre-monsoon drizzles. In the case of red gram and castor, the seeds are placed in the centre of the furrow and covered with finely-powdered farmyard manure. In the cases of sorghum, pearl millet, bengal gram, sunflower, safflower *etc.,* the seeds are placed in the soil using a seed drill. Dry seeding is practiced when delayed rain is forecast. However, the seeds should not be placed in partially moist soil before the commencement of sowing rains.

Intercultural operations: The weeds should be kept under control by herbicides, inter cultivation as well as hand weeding within 3–4 weeks from seeding. Frequent inter cultivation (2–3 times) using sweep-type hoes controls weeds and checks evaporation. In addition, inter cultivation also provides aeration and improves infiltration of rain-water into the soil. To control evaporation from the soil surface, various types of mulches should be planned in wide-row crops especially during *rabi* season. Organic mulches of crop residues and weeds removed before seed formation is extremely useful as they are converted to organic manures on decomposition later in the year or in subsequent years.

Disease and pest management: In rainfed farming, our approach is to adopt suitable agronomic practices to avoid pests rather than the use of costly chemicals. Early sowing leads to vigorous seedlings, which can easily withstand the knockout problems. Crop rotation helps in reducing the incidence of *Striga* in jowar and *Orobanche* in tobacco. Border crop of jowar/maize is helpful in controlling necrosis in sunflower.

xii. Inter–terrace management practices

Large number of *in-situ* water conservation practices has been tested over years under different agro-ecological situations. Following are generally recommended both in *Alfisols* and *Vertisols* watersheds. Some are for increasing the infiltration and others for smooth disposal of surplus water.

1. Land smoothing
2. Early tillage
3. Graded ridges
4. Furrow on grade
5. Contour cultivation/contour bunds
6. Inter-cultivation
7. Increasing infiltration
8. Surface mulches
9. Vertical mulching
10. Compartment bunds
11. Scooping

xiii. Wind Break or Shelter belt

To minimize the erosion hazards of speedy winds and optimize production of agricultural crops, various efforts have been made in the past by adopting different soil conservation measures. Among them the shelterbelt plantations rank high due to its expected effects on reduction in wind velocity. Shelterbelts are the vegetative barriers that are designed to reduce wind speed and provide sheltered area on leeward side. They can decrease wind speed to a level below the threshold value required for lifting and movement of soil particles and hence, are expected to reduce impact of strong winds. The reduction in wind velocity leads to reduction in soil loss, deposition of sand on road, canal and other water bodies. The work on shelterbelt was started way back in 1950s including design of shelterbelt, screening of suitable species for its composition and developed techniques for plantation establishment. The shelterbelts also reduces ground temperature, possible weight loss in animals and improves the crop yield.

Choice of plant species for a shelterbelt

The selection of a plant species for a shelterbelt depends on climatic and edaphic conditions of the regions. Shelterbelts selected should have quick spreading growth. The plant species selected for shelterbelts are given in Table 50.

Table 50: Trees and shrubs for a shelterbelt

S. No.	Trees	Area recommended
1.	*Acacia tortilis, Eucalyptus camaldulensis, Dalbergia sisso, Tecomella undulata and Dalbergia sisso*	Jaisalmer, Rajasthan
2.	*Acacia albida, Albizia* spp., *Anacardium occidentale, Annona senegalensis, Azadirachta indica, Balanites aegyptiaca, Calliandra calothyrsus, Calodendrum capense, Cassia siamea, Casuarina* spp., *Cupressus lusitanica, Ekebergia capensis, Eriobotrya japonica, Eucalyptus* spp., *Gliricidia sepium, Gmelina arborea, Grevillea robusta, Hakea saligna, Juniperus procera, Macadamia tetraphylla, Mangifera indica, Markhamia lutea, Morus alba, Olea europaea, Prosopis* spp., *Prunus africanus, Psidium guajava, Spathodea campanulata, Syzygium cuminii, Trichilia emetica, Vitex* spp., *Ziziphus spp*	Trans Nzoia, Nakuru, and Uasin Gishu, Kenya
3.	*Casuarina* spp., *Azadirchta indica, Cassia* spp., *Prosopis* spp., *Leucaena spp.* and *Agave* spp.	Coimbatore, Tamilnadu
4.	*Cilicia* spp., *Terminalia* spp., *Careya arborea, Aporosa lindleyana, Vateria indica, Leucaena leucocephala* and *Casuarina equisetifolia*	Uttara Kannada, Karnataka
5.	*Cassia, Prosopis, Leucaena, Casuarina, Acacia, Grevillea, Syzygium, Dalbergia, Saccharum spontaneum, S. munja, Panicum antidotale, Cencnrus* sp, *Calotropis procera, Clerodendron phlomoides, Cassia auriculata, Dodonaia viscosa, Acacia arabica, A. leucopholea, Dalbergia sissoo, Lannea coromandelica, Eucalyptus* spp., *Tamarix, articulata, Parkinsonia aculeata, Prosopis juliflora, Prosopis spicigera, Casuarina equisetifolia.*	Dry land regions of India

Growing of Shelterbelts in crop lands helps to reduce the moisture evaporation from the soil (Table 51).

Table 51: Effect of *Cassia siamea* shelterbelt on pan evaporation

Location	Pan evaporation (mm day^{-1})			
	April	May	June	July
Windward side	11.0	14.5	13.4	6.0
Leeward side	9.5	13.4	12.5	5.7
% decrease	14	8	8	5

Source: Anon., 2006

xiv. Other agronomic practices: like broad bed and furrow system, wider row spacing with deep intercultivation, sowing across slope, high density planting through multistoreyed, set row cultivation, gravel, and sand mulching and staggered planting.

Wider row spacing and frequent deep inter-cultivation: Adoption of wider row spacing ensures larger cross section of effective root zone, which provides moisture for longer period. Frequent deep intercultivation reduces evaporation considerably by creating dust mulch. This wider row also helps for better light interception; crops make use of the space between the rows. It is very much suited to the crops which are more exhaustive for nutrients and moisture.

Wider row spacing with frequent deep intercultivation with blade harrow increased the yield of pearl millet to 500 kg per ha as against 250 kg per ha in case of farmers practice. The sunflower yield was 600 to 750 kg per ha with wider row spacing as against an yield of 250 to 300 kg per ha under normal spacing 36–40 cm. The effectiveness of this technology was more observed during sub normal rainfall years (Table 52).

Table 52: Wider row spacing with Frequent deep intercultivation

Crop	Yield (kg ha^{-1})		% increase over farmers practice
	Wider row spacing	Farmers practice	
Pearl millet	500	250	50
Sunflower	675	275	41

Source: Ramachandrappa *et al.* (2014)

Set row cultivation: is suitable for medium to deep black soil for groundnut and cotton crops. Before opening of set rows, deep ploughing is done to break the hard pan in profile. Then land is harrowed to bring the land to fine tilth. Set furrows are opened according to spacing followed for the crops. The well decomposed FYM @ 5–6 t ha^{-1}, green manure crops *viz.* gliricidia and crop residues are filled in the set furrows. Further the crops are sown in the set rows after receipt of rain. This type of cultivation helps to increase the efficient use of nutrients, moisture and reduce the cost of cultivation and increases the biological properties of soil (Table 53).

Table 53: Yield of crops as influenced by set row cultivation

Crop	Yield (kg ha^{-1})		% increase over farmers practice
	Set furrow cultivation	Farmers practice	
Pearl millet (*Kharif*)	1092	749	69
Sunflower (*Rabi*)	457	295	65
Pigeonpea	1112	732	52

Source: Ramachandrappa *et al.* (2014)

Gravel and sand mulching

Sodic medium and deep black soils exist an extent of 2.5 lakh ha in Koppal and Gadag districts in northern Karnataka. The infiltration rate is low and most of the rainwater is lost as runoff in these soils. However, traditionally, some farmers in this region apply locally available gravel and sand mixture as mulch and successfully produce better crops and get more income. The technology involves sand application during summer to sodic *vertisols*. Before application of sand, perennial weeds like *Cynodon dactylon*, *Cyperus rotundus etc.*, are removed. FYM @ 5 t ha^{-1} is applied followed by deep ploughing. After bringing the soil to fine tilth, nearly 275–300 tractor loads ha^{-1} of gravel and sand mixture is uniformly applied and spread manually using spades to ensure uniform thickness of 10 to 15 cm on soil surface. This is a capital investment technology of land development and mulching but is effective for 15–20 years. However, gravel sand should be applied once in 5 years in patches where sand is lost due to runoff. Gravel sand mulching helps in conserving rainwater *in-situ*. The technology is adopted in Bijapur, Bagalkot, Gadag, Koppal, Bellary part of Dharwad, Belgaum, Raichur and Davangere districts in medium to deep black soils of Northern dry zone of Karnataka.

This practice enables double cropping of groundnut or greengram in *kharif,* followed by *rabi* sorghum or sunflower or chickpea in *rabi*. Further, this practice gives a yield advantage up to 125% in groundnut (1400 kg ha^{-1}), 214% in *rabi* sorghum (2200 kg ha^{-1}), 300% in sunflower (2000 kg ha^{-1}), 300% in chickpea (1500 kg ha^{-1}) and 366% in greengram(1400 kg ha^{-1}) with additional net returns from ₹5,300 to ₹33,000 and B:C ratio of 2.83 to 5.69. The cost of sand application can be recovered within 2 years of cropping (Table 54).

Table 54: Yield of crops as influenced by Gravel sand mulching

Crop	Yield (kg ha^{-1})		% increase over farmers practice
	Gravel sand mulching	Farmers practice	
Greengram	800–1400	200–300	300–366
Rabi Sorghum	1400–2200	600–700	133–214
Sunflower	1200–2000	400–500	200–300
Chickpea	1200–1500	300–400	275–300
Groundnut	1500–1800	700–800	114–125

Source: Ramachandrappa *et al.* (2014)

2. Mechanical methods of soil moisture conservation: The basic principle involved are:

- shaping the land surface manually or with implements to reduce the velocity of runoff,
- to allow more time for rainfall to stand on soil surface and
- to facilitate more infiltration of rainfall into soil layers

Choice of any particular method under a given situation is influenced by rainfall characters, soil type, crops, sowing methods and slope of land given in Table 55 and 56.

Table 55: Mechanical structure under dry land conditions

Measures	Annual rainfall (mm)	Slope of land (%)	Soil Type
Leveling and Grading	No upper limit	Less than 3	Medium deep soils
Contour bunds	700	1 to 6	All permeable soils
Graded bunds	800	1 to 6	All types

Bench Terracing	1100	Any slope	Soil suitable for excavation
Contour trenching	1200	Any slope	Soils suitable for excavation

(i) Basin listing/scooping: Formation of small depressions (basins) of 10–15 cm depth and 10–15 cm width at regular intervals using an implement called basin lister. The small basins collect rainfall and improve its storage. It is usually done before sowing. It is suitable for all soil types and crops.

(ii) Bunding: Formation of narrow based or broad based bunds across slope at suitable intervals depending on slope of field. The bunds check the free flow of runoff water, impound the rainwater in the inter-bund space, increase its infiltration and improve soil moisture storage. Leveling of inter-bund space is essential to ensure uniform spread of water and avoid water stagnation in patches. It can be classified into three types:

Table 56: Choice of bunds for different soil and rainfall situations

Type of bunds	Rainfall region	Remarks
Contour bunds	Low rainfall region (< 750 mm)	Suitable for medium soils of slope upto 6%. Not suitable for *Vertisols*
Graded bunds	Medium to high rainfall regions (> 750 mm)	Suitable for *Vertisols* when provided with up-stream channel. With Alfisols, no upstream channel needed
Graded border strips	Medium rainfall regions (750–1150 mm)	Suitable for deep *Alfisols* and Vertic *Inceptisols*

a. Contour bunding: Bunds constructed along the contours or with permissible deviation from contours are called contour bunds. Bunding is preferred over the areas where annual rainfall is less than 500 mm and soil moisture is limiting for crop production. Mostly, bunds are constructed on the land slope varying from 2 to 10%. For the lands below 2%, agronomic measures can be adopted for *in-situ* moisture conservation. Contour bunds are preferred in the permeable soils like alluvial, red and shallow black soils. Bunds of 1 m basal width, 0.5 m top width and 0.5 m height are formed along the contour (across the slope). The distance between two contour bunds depends on slope. The inter-bund surface is leveled and used for cropping. The criterion for spacing of contour bunds is to intercept the water before it attains erosive velocity. This depends on slope, soils, rainfall, cropping programme and conservation practices adopted. It is not suitable for heavy black soils with low infiltration where bunds tend to develop cracks on drying. Contour bunds are permanent structures and requires technical assistance and heavy investment.

The maintenance of bund is very important for better conservation. Planting of useful grasses and periodical strengthening of bunds at weak points are important maintenance practices.

Construction of contour bunds and their distance largely depend on the velocity of running water and slope of the land. They can be made by manual labour and or by animal or mechanically operated machines. Bulldozers and animal drawn scrappers can be employed for contour bunding works depending upon the size of the bunds. The specifications of constructing contour bunds are given in the Table 57.

A series of such contour bunds divides the area into strips and acts as barriers to the flow of water. It allows the water to percolate and conserve the soil. Spacing between the bunds is an important factor that decides the efficiency of functioning of contour bunds and they have to be placed for intercepting the erosive velocity.

Spacing between the bunds should also not hinder different agricultural operations. It is usually expressed in terms of vertical interval (difference in elevation between two similar points on two consecutive bunds). The general recommendation spacing between the contour bunds is given in Table 57.

Maintenance of contour bunds

Contour bunds should be periodically attended the maintenance works for better efficiency and prolonged benefits. Lack of sufficient measures may reduce the efficiency of bunds and degradation may bring more damage. The different maintenance works include the following.

1. Plugging the rat and rabbit burrows and breaches
2. Filling up of sags in bunds
3. Maintenance of bund height and section
4. Repairing the outlets
5. Growing grasses and trees on bunds

Table 57: Contour bund specifications

S. No	Depth of Soil	Base width (m)	Top width (m)	Height (m)	Side slope
1.	Shallow soil (7.5–22.5 cm)	2.67	0.38	0.75	1.5:1
2.	Medium soils (22.5–45.0 cm)	3.12	0.60	0.85	1.5:1
3.	Medium deep soil (45.0–90.0 cm)	4.25	0.60	0.90	2.0:1

Table 58: Spacing between contour bunds

Slope of the land (%)	Vertical interval (m)	Horizontal distance (m)
0–1	1.05–1.20	>100
1–2	1.20–1.50	75–100
2–3	1.50–1.65	60–75
3–5	1.65–1.80	40–50
5–6	1.80–1.95	<40

b. Graded/field bunding: Bunds of 30–45 cm basal width, and 15–20 cm height are formed across slope at suitable intervals of 20–30 m depending on slope. The inter-bund area is leveled and cropped. It is suitable for medium deep to deep red soils with slopes up to 2%. It is not suitable for black soils due to susceptibility to cracking and breaching. Bunds can be maintained for 2–3 seasons with reshaping as and when required.

Graded bunds consist of small bunds constructed with a slope of 0.3 to 0.5% in order to dispose the excess water through the graded channels which leads to a naturally depressed area of the land. The purpose of the bund is not to store the excess water, but to retain it in farm ponds through graded channels. They are recommended for high rainfall areas (more than 600 mm per year) and generally used for the safety and areas of impervious soil. They are also practicable in areas receiving an annual rainfall of 500 mm in highly impermeable soils (black soils). The design of disposal channel should be within the eorrosive limit. Graded bunds can be constructed with manual labour or with animal drawn scrapper or bulldozer.

The soil for the bund may be collected from either side of the channel or by scraping from either side of the bund if no channels are required. Top width should be increased in proportion with height of the bund. Suitable spillovers are required for the purpose of safe removal of excess water. In soils of high permeability, biological measures can be combined and the interval between the bund can be increased.

Graded bunds need to be inspected after heavy rains for the purpose of seeing and rectifying the depressions, rill formations and undue settlements in the bunds. After care practices like leveling, cutting or filling may be done to eliminate the defects. Graded bunds check the runoff water, reduce the soil loss and increase the moisture status (Table 59). Graded border strips are usually long, uniformly graded strips of land, separated by earth bunds (Fig. 15).

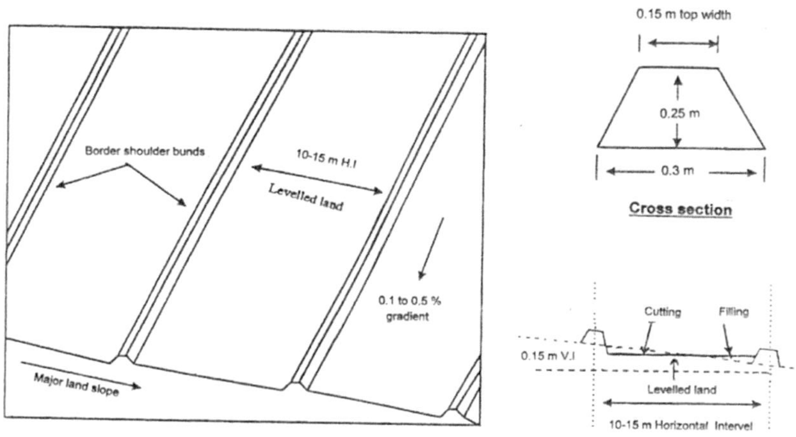

Fig. 15. Cross sectional view of graded border strips

Table 59: Effect of conservation measures on runoff, soil loss and crop yields in red soil areas

Land treatment	Runoff of annual rainfall (%)	Soil loss t ha^{-1}	Soil moisture at 0–15 cm depth in different std. weeks				Grain yield of finger millet q ha^{-1}
			46	**47**	**48**	**49**	
Control (along the slope cultn.)	26.30	11.01	07.69	3.85	3.68	3.35	18.60
Existing bunds (cultn. across slope)	20.10	07.36	09.09	6.49	3.74	3.62	19.30
Graded bunds (cultn. parallel to bunds)	10.62	03.71	10.13	7.59	4.58	3.81	24.12
Contour bunds	03.80	01.30	11.84	8.00	5.76	4.32	24.25
Khus on contour	07.90	02.48	11.39	7.85	5.68	4.18	24.75

Source: Krishnappa et al. (1994)

Advantages of graded bunds

1. Safe and efficient drain for excess water
2. Elimination of water stagnation
3. Involvement of less cost (as compared to contour bunds)
4. For other agricultural practices like strip cropping, contour farming *etc.*

Drawbacks with graded bunds

1. Need of grassed water for protection
2. Possible weed spread by drained water
3. Water removed or drained once from the field cannot be brought back (unless stored in farm ponds)

c. Terracing

A terrace is a ridge or embankment of earth constructed across the slope to control run off and minimize soil erosion. It reduces sheet and rill erosion and prevents gully erosion. It will be one of the expensive measures in cropped lands and it is most suited in hill slopes with supplementary conservation measures like contour tillage, crop rotation and strip cropping.

The different types of terraces are

i. Broad based terraces: They are also called as ridge terraces or broad bunds. The terrace consists of a channel with a ridge on its lower and whose function is to drain surface runoff and encourage infiltration. Thus, these are also called as drainage or absorption terrace. These can be adopted in slopes < 10%.

ii. Bench terraces: Bench terraces are preferred in the land slopes varying from 10 to 33% on hill slopes. The cost is a limiting factor in adoption of bench terraces. They are adopted where intensive cultivation is practiced on hill slopes. Bench terraces consist of constructing step like fields and contours by half cutting and half filling. Original slope is converted into level fields eliminating all hazards of soil erosion.

There are two types of terracing namely bench terracing and broad base ridge terrace.The details of these terraces and their construction are explained in Table 60.

Table 60: Construction of bench terracing and broad base terracing

Type of terrace	Rainfall/ Area	Slope (%)	Construction meth-od	Uses
Bench terracing	1100	10 to 33 (at times 5 to 10)	Step line fields along contours and length may be 400 to 500 m	Many crops can be grown (even rice in hill slopes)
Broad base terracing				
a. Graded	Humid areas	1 to 15	Terracing at constant or varible grade along its length to convey excess water Vertical Spacing: 0.75 to 2.85 m Horizontal spacing: 75 to 90	
b. Level	Semi arid condition	1 to 3	Terracing on absoulate contour length of 120 to 150	Increases infiltration and absorption

iii. Zingg conservation bench terraces: In this practice, 1/3 to 1/4 area on the upstream side of the bund are levelled (Fig. 16). The rainwater is allowed to spread uniformly along the bund in levelled portion. The pressure on the bund is reduced. The bunds have to be raised at every 4–5 years interval. It is suited for black soil areas.

Double cropping was made possible due to this modified practice in the farmers fields. The extent of increase in the yield of different crops ranged between 39 to 139%. Adoption of this technology enhanced the productivity of *rabi* sorghum, sunflower, greengram and pearl millet by 39, 82, 113 and 139%, respectively over a period of 5 years (Table 61). The yield advantage is more visible during deficit rainfall period.

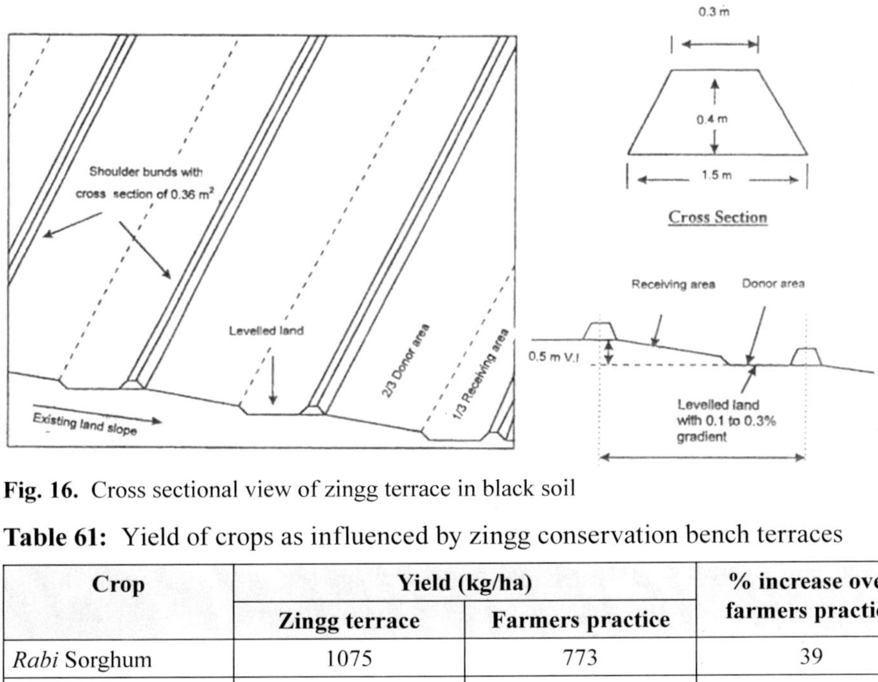

Fig. 16. Cross sectional view of zingg terrace in black soil

Table 61: Yield of crops as influenced by zingg conservation bench terraces

Crop	Yield (kg/ha)		% increase over farmers practice
	Zingg terrace	Farmers practice	
Rabi Sorghum	1075	773	39
Sunflower	1220	670	82
Greengram	320	150	113
Pearl millet	550	230	139

Source: Ramachandrappa *et al.* (2014)

iv. Level or table top bench terraces: Bench terraces with level tops are adopted in the medium rainfall, having deep and highly permeable soils. As there is no slope given to the benches, most of the rainfall falling in the area is to be absorbed by the soil and very little water goes as surface drainage. Terraces can be used for paddy cultivation where irrigation facility is available.

v. Inward sloping bench terraces: Terraces with benches sloping inward are adopted in the high rainfall areas where a major portion of the rainfall is to be drained as surface runoff. In such terraces, a suitable outlet at the inward end of each of these terraces is to be provided to remove excess runoff.

vi. Outward sloping bench terraces: Bench terraces with outward sloping are adopted in low rainfall areas with permeable soils. In these terraces, a strong bund with spillway arrangement would remove excess runoff occurring in case of heavy rainfall events. Excess runoff will flow one terrace to another.

vii. Stone terraces: They are small embankments constructed with stones across the hill slopes. These are justified where stones are available in adequate quantities near to the site. These can be used for growing annual crops and perennial tree plantation.

d. Compartmental bunding: Small bunds of 15 cm width and 15 cm height are formed in both directions (along and across slope) to divide the field into small basins or compartments. It is suitable for red soils and black soils with a slope of 0.5 to 1%. The bunds can be formed before sowing or immediately after sowing with local wooden plough. It is highly suitable for broadcast sown crops.

In northern dry zone of Karnataka, *kharif* cropping is not possible due to workability and tillage related constraints in medium to deep black soils. Further, infiltration rate is low resulting in more runoff. It involves making square compartments on the field to retain rainwater and arrest soil erosion. After receipt of early rains in June and July, land is harrowed to remove germinating weeds. The compartmental bunds (0.15 m height) are formed using bullock drawn bund former. The size of the bunds varies from 3 m × 3 m–4.5 m × 4.5 m depending on the slope. These bunds retained till the sowing of *rabi* crops, which are sown with seed cum fertilizer drill during second fortnight of September to first fortnight of October.

Compartmental bunds provide more opportunity time for water to infiltrate into the soil and help in conserving soil moisture. The technology is popularised in the Operational research project and adopted in the districts of Bijapur, Bagalkot, Gadag, Koppal Bellary, part of Dharwad, Belgaum, Raichur and

Davangere districts in medium to deep black soils of Northern dry zone of Karnataka.

Rabi sorghum, sunflower, safflower and chickpea gave seed yield of 870, 675, 620 and 450 kg ha^{-1}, with a yield advantage of 40, 35, 38 and 50% respectively with additional net returns of ₹2,475, ₹3,700, ₹3,250 and ₹2,850 with compartment bunding over flat planting. The impact of the practice is more during sub-optimal rainfall years. It also significantly controls run off.

e. Rubble bunds

Small stone sand pebbles locally available can be used to erect bund all along the contour or across the slope. So this will resist runoff and help in better moisture conservation.

f. Ridges and furrows

Ridges and furrows formed with 0.2 to 0.4% grade as an inter terrace management will be useful in conserving moisture for widely spaced crops like cotton, maize, sorghum, castor and red gram.

Furrows of 30–45 cm width and 15–20 cm height are formed across slope. The furrows guide runoff water safely when rainfall intensity is high and avoid water stagnation. They collect and store water when rainfall intensity is less. It is suitable for medium deep to deep black soils and deep red soils. It can be practised in wide row spaced crops like cotton, maize, chillies, tomato *etc.* It is not suitable for shallow red soils, shallow black soils and sandy/gravelly soils, broadcasted crops and for crops sown at closer row spacing (<30 cm). Since furrows are formed usually before sowing, sowing by dibbling or planting alone is possible.

Ridges and furrows help in conserving soil moisture. The rainwater is conserved in the furrows where it falls as the ridges and furrows provide more opportunity time for water to infiltrate into the soil. Adoption of ridges and furrows in *rabi* sorghum, sunflower, safflower and chickpea resulted in yield advantage of 51, 38, 46 and 53%, respectively over control or flat planting (Table 62).

Table 62: Performance of crops under ridges and furrows

Crop	Yield (kg ha^{-1})		% increase yield over farmers practice
	Ridges and furrows	Farmers practice	
Rabi Sorghum	953	633	51
Sunflower	725	525	38
Safflower	670	460	46
Chickpea	750	490	53

Source: Ramachandrappa *et al.* (2014)

Tied ridging is a modification of the above system of ridges and furrows where in the ridges are connected or tied by a small bund at definite intervals (with a minimum of three meters) and they will act as a barrier. The furrows act as the storage space for the rainwater. These tied ridges will enhance the availability of soil moisture to the crops. It is highly advantageous in dry conditions as compared to high rainfall periods or areas. For the purpose of avoiding any damage to tied ridges, the furrows can be constructed on gentle slope to favour run off. The infiltration rate of soil can be increased with mulching and subsoiling (Table 63 & 64).

Random tie ridging is another modification where discontinuous furrows of 20–25 cm width, 45–60 cm length and 15 cm depth are formed between clumps or hills of crops at the time of weeding.

Yet another modification of ridges and furrows method is the practice of sowing in lines on flat beds and formation of furrows between crop rows at 25–30 days after sowing. This enables sowing behind plough or through seed drill.

Table 63: Performance of tied ridges in black soils

Treatments	Pigeonpea seed yield (kg ha^{-1})	% increase over farmers' practice	*Rabi* sorghum grain yield (kg ha^{-1})	% increase over farmers' practice
Tied ridging	1344	17.6	953	51
Farmers' practice	1143	–	633	–

Source: Ramachandrappa *et al.* (2014)

Table 64: *In situ* moisture conservation through tied ridges and mulching in chilli varieties

Variety	DOT	Duration (Days)	Dry fruit yield (kg ha^{-1})		% increase in yield	RWUE (kg ha–mm^{-1})	Net returns (₹ha^{-1})	B: C ratio
			With mulch	Without mulch				
Samrudhi	24th Aug	202	**1359**	960	42	2.12	132049	5.26
Chikkaballapur local	24th Aug	202	**1104**	776	42	1.72	101449	4.27

Source: Ramachandrappa *et al.* (2014)

Mulching

Without mulching

g. Broad bed and furrow (BBF)

In semi arid tropics, high intensity rains happen at times of dry season or early season rain fed crop when crop cover is limited. This leads to severe soil erosion and run off in deep vertisols, which may not provide adequate surface drainage. Thus, there is a need for an *in-situ* moisture and soil conservation and drainage technology in such areas. A system called broad bed and furrow developed by International Crop Research Institute for Semi Arid Tropics center has been found satisfactory in deep vertisols especially under conditions of high and assured rainfall. The system consists of a relatively flat bed and ridge approximately 90–150 cm wide (depending upon soil type) and a shallow furrow of about 30 cm wide and 15 cm depth. Furrows should be made only after assessing the directions of cultivation and topography of the land. A multipurpose tool carrier called tropicultor, which has two ridges, can be used for this purpose. It is important to have the ridges operate at shallow depth in order to attain straight lines. A bed former can then be used to shape the beds. The topography of deep vertisols where broad bed furrow system found suitable is 0.4–0.8 per cent slopes. The length of the furrows should be 100 m or less to discharge excess water. It is suitable for heavy black soils and deep red soils.

Advantages of broad bed furrows

1. It provides great flexibility to fit crops and cropping systems
2. Precision placement of seeds and fertilizers are possible
3. The system can also be used for supplemental irrigation
4. The seed bed at the time of beginning of cropping season will be receptive to rainfall favouring good storage in soil surface layers
5. Broad bed furrows provide better drainage and protect the soil from erosion through out the season
6. It facilitates easy sowing in raised beds of friable soil

Precautions in making broad bed furrows

1. It is important to obtain uniform shape without sharp edges for adopting different crops and cropping systems
2. Subsequent cultivation may be needed to eliminate weeds and to shape up the broad bed furrows
3. Soil compaction due to animals or implements should be limited to furrows

h. Dead furrow: At the time of sowing or immediately after sowing, deep furrows of 20 cm depth are formed at an interval of 6–8 rows of crops. No crop is raised in the furrow. Sowing and furrowing are done across slope. It can be done with wooden plough in both black and red soils.

Groundnut + Pigeonpea (10:2) with staggered moisture conservation furrow

Table 65: Yield of finger millet + Pigeonpea (8:2) cropping system as influenced by soil and moisture conservation practices

Crop	Duration	Grain yield (kg ha^{-1})		% increase in yield	RWUE (kg ha-mm^{-1})	Net returns (₹ ha^{-1})	B:C ratio
		With *In-situ* moisture conservation practice	With normal practice (Akkadi)				
Finger millet + pigeonpea (8:2)	16th Jul to 26th Dec	3056	2100	45	4.71	52586	3.12

Inter cropping of finger millet (MR-1) + pigeonpea (BRG–2) in 8:2 with conservation furrow between paired rows of pigeonpea recorded higher finger millet grain equivalent yield, net returns and B: C ratio (3056 kg ha^{-1}, ₹52,586 kg ha^{-1} and 3.12, respectively) compared to farmers practice of Finger millet + Akkadi cropping with an yield increase of 45%. Similarly in Groundnut pigeon pea (8:2) intercropping system dead furrow or conservation furrow is formed between two pigeon pea lines (Table 65).

i. Deep trencher for soil moisture conservation

Deep trencher helps in opening the trenches of depth up to 30–60 cm and a width of 50–60 cm. Trenches opened by using tractor drawn deep trencher was found to be suitable for conserving runoff water as compared to other implements. The less runoff (mm), highest fruit yield and highest soil moisture was found in orchards. The rain water use efficiency was found to be highest in deep trencher. Opening of trench in between the mango (orchards) tree would be helpful in getting higher yield by conserving soil moisture through rain water harvesting in the deep trench and also lasts more than two years.

j. Micro catchments: A micro catchments is a specially contoured area with slopes and berms designed to increase runoff from rain and concentrate it in a planting basin where it infiltrate and is effectively "stored" in the soil profile. The water is available to plants but protected from evaporation.

a. Circular basin of one meter diameter for lands depending upon infiltration and rainfall

b. 'V' ditches of size 5 m × 5 m with trees planted centre and height according to the rainfall and slope of sloppy lands

c. Saucer basins/Semi circular bunds with 2 m diameter to height of 15–20 cm across the slope

k. Catch Pits

In this method pits of size 1 m × 1 m × 1 m are dug at many places in the corners of the fields or at any suitable place in slope areas. At the time of monsoon, the run off water is collected in the pits along with the silt. This will enhance the moisture availability to the crops.

3. Agrostological methods: Refers to the use of grasses to control soil erosion, reduce run off and improve soil moisture storage. Grasses with their close canopy cover over soil surface and profuse root system which binds soil particles, provide excellent protection against runoff and erosion. The following are the various agrostological methods of *in-situ* moisture conservation.

i. **Pasture/grass lands:** Raising perennial grasses to establish pastures/grass land is recommended for shallow gravelly, eroded, degraded soils. Grass canopy intercepts rainfall, reduces splash erosion, checks runoff and improves soil moisture storage from rainfall.

ii. **Strip cropping with grasses:** Alternate strips of grasses and annual field crops arranged across slope check runoff and erosion and help in increasing moisture storage in soil.

iii. **Ley farming:** is the practice of growing fodder grasses and legumes and annual crops in rotation. Grasses and legumes like Cenchrus, stylo are grown for 3–5 years and followed by annual crops like sorghum for 2 year.
iv. **Vegetative barriers:** consists of one or two rows of perennial grasses established at suitable interval across the slope. It serves as a barrier for runoff and soil transport. Vetiver, Cenchrus *etc*., are suitable grasses for this purpose.

Vetiver can be planted in rows at intervals of 40 m in 0.5% slope. Plough furrows are opened with disc plough first before commencement of monsoon. 5–8 cm deep holes are formed at 20 cm interval and two slips per hole are planted in the beginning of rainy season. The soil around the roots is compacted. Vetiver barriers check runoff and prevent soil erosion. While they retain the soil they allow excess runoff to flow through their canopy without soil loss. It is adapted to drought and requires less care for maintenance. It does not exhibit any border effect on crops in adjacent rows. Use of vegetative barriers for conserving soil and moisture in agricultural lands is inherent with Indian agriculture. Vegetative barriers are hedgerows of perennial grasses or shrubs planted at a regular interval for conserving soil and water in slope lands. They have great potential in arresting soil and run off losses thereby increasing the time of absorption. The efficiency of moisture conservation depends on the hedge forming ability of the plant species. These barriers are easy to raise, less labour intensive and acceptable to the farmers. Vegetative barriers can be planted with or without earthern embankments. The selection of the species is based on the morphological and physiological characteristics that are ideal for the purpose of conserving soil and moisture.

The selected plant species should have the following characteristics:

1. Local adaptability
2. Native to the region or zone
3. Fast growing in nature
4. Should be perennial with stiff and erect stems
5. Ability to form dense hedge when planted on rows

Planting grasses like khus and nase on bunds will reduce soil loss by erosion and holds moisture for longer period which facilitates crop to survive in dry periods. By adopting live bunds in field with khus or nase grass will help in reducing the soil loss and runoff by 36% in slopes upto 2.5%. Live barriers reduce runoff (11–12%) & soil (1.52–1.70 t ha^{-1} $year^{-1}$) as compared to control which results in higher soil loss (2.5 t ha^{-1} $year^{-1}$). Khus and Nase grasses help in reducing soil loss and decreases silt accumulation in reservoirs, nalas, *etc.* The most suitable plant species of vegetative barriers under dry land conditions (Table 66).

Table 66: Vegetative barriers for dry lands

S. No.	Common name	Botanical name
1.	Khus or Vettiver	*Vettiveria zizinoides*
2.	Bermuda grass	*Cyanodon dactylon*
3.	Buffel grass	*Cenchrus ciliaris*
4.	Bird wood grass	*Cenchrus setigerus*
5.	Lemon grass	*Cymbopogon jwarancusa*

Vegetative Barrier

Khus-live barrier

Laucaena - Live barrier

Agave

Vegetative barriers are best suited for black soil. Unlike contour bunding which give way to development of crack in summer in black soils vegetative barriers do not allow such phenomenon (Table 67).

Table 67: Yield of finger millet as influenced by inter terrace management practices

Treatments	Yield (kg ha^{-1})		HI	B:C	RWUE (kg ha–mm^{-1})
	Grain	Straw			
Khus grass Upper	2300	3375	0.41	2.56	4.6
Lower	2475	3575	0.41	2.76	5.0
Mean	2388	3475	0.41	2.66	4.8
Nase grass Upper	2550	3650	0.41	2.84	5.1
Lower	2725	3925	0.41	3.03	5.5
Mean	2638	3788	0.41	2.94	5.3
Control	1925	2350	0.45	2.13	3.9

Source: AICRPDA

Soil and Moisture Conservation for Hilly and Coastal Regions

In Hilly areas the major problem is steep slope susceptible to erosion causing severe soil loss and problem of retaining water and nutrients. It is well accepted that converting the land configuration into terraces with appropriate watershed models for the diversion of excess water and storage for longer period apart from appropriate crop and variety depending on the situation (upland, midland and lowland) would go a long way in conserving natural resources and ensuring higher productivity.

Coastal ecosystem is plagued with porous laterite soils which are highly pervious resulting in substantial leaching of plant nutrients which finally enter into seashore affecting the aquatic life apart from soil erosion. The dominant small holdings in both hilly and coastal areas demand appropriate farming system apart from conservation of soil and nutrients. Integrated farming system modules based on the bio-physical resources of the farmers would provide food, nutrition and livelihood security of farmers. Integrated farming system can include paddy, fish/prawn cultivation, multi-storied plantation based cropping system involving coconut, areca nut, cardamom, pepper, banana, pineapple, mango, sapota *etc.* It is also possible to increase the cropping intensity with short duration pulses in paddy fallows.

The problem of salt affected soils is wide spread in irrigated ecosystems of the state limiting crop production seriously due to unfavourable physical, chemical and biological properties. Successful reclamation of sodic soils includes land levelling, bunding, addition of amendment, overcoming nutrient deficiencies, choice of tolerant crop and variety, proper water management *etc.* Provision for drainage can help to ameliorate saline soils (Yadav, 2008). Soil acidity is another problem in Hilly and coastal areas need application of lime and organic matter to restore.

Chapter 7

Drought Management Strategies

Improving water use efficiency or enhancing agricultural water productivity is a critical response to growing water scarcity, including the need to leave enough water in rivers and lakes to sustain ecosystems and to meet the growing demands of cities and industries. Rainfall is the main source of water for the existence of life on the earth. Timely onset of rainfall and its distribution during the crop growing season determine the success of agriculture and food production. Unfortunately, the human life has hardly any control over nature and prevailance of delayed onset of monsoon and dryspells during different stages of crop growth are commonly observed in the dryland agriculture. Drought is a weather-related natural disaster which makes greater impact on food production and it reduces life expectancy and the economic performance of large regions or entire countries. Drought is a recurrent feature of the climate. It occurs in virtually all climatic zones, and its characteristics vary significantly among regions. Drought differs from aridity in that drought is temporary; aridity is permanent characteristic of regions with low rainfall. Drought is related to a deficiency of precipitation over an extended period of time, usually for a season or more. This deficiency results in a water shortage for some activity, group or environmental sector. Drought is also related to the timing of precipitation. Other climatic factors such as high temperature, high wind and low relative humidity are often associated with drought.

Drought may occur due to

- Deficit of rainfall
- Depletion of ground water
- Inadequate supply of water through surface water bodies like rivers, streams, canals, tanks *etc*.
- Insufficient moisture in the sub surface layer of the soil and
- Increase of demand for water as compared to availability

Drought conditions may lead to

- Decline in food and fodder production
- Fall in industrial production
- Distress sale of cattle

- Migration of people
- Unemployment and
- Poverty

Effect on agriculture production during subsequent years
- Non availability of quality seed for production
- Inadequate draft power for agricultural operation due to distress sale of animals
- Reduced use of inputs like fertilizers due to decline in investment capacity of farmer
- Non availability of raw material for agro based industries
- Deforestation to meet the energy needs in domestic sector

Rainfall conditions are defined as follows
- Excess: + 20% or more of the average of 70–100 years
- Normal: + 19% to –19% of the average of 70–100 years
- Deficient: –20% to –59% of the average of 70–100 years
- Scanty: –60% or less of the average of 70–100 years

Thornthwaite (1947) defined drought as a "**condition in which the amount of water needed for transpiration and direct evaporation exceeds the amount available in the soil.**
- Moderate drought: seasonal rainfall deficiency is 26–50% of the normal
- Severe drought: seasonal rainfall deficiency is > 50% of the normal
- Drought year: Less than 75% of the normal rainfall

Drought definitions are of two types: (1) Conceptual (2) Operational
Conceptual definitions help to understand the meaning of drought and its effects. For example, drought is a protracted period of deficient precipitation which causes extensive damage to crops, resulting in loss of yield.

Operational definitions help to identify the drought's beginning, end and degree of severity. To determine the beginning of drought, operational definitions specify the degree of departure from the precipitation average over some time period. This is usually accomplished by comparing the current situation with the historical average. The threshold identified as the beginning of a drought (*e.g.*, 75% of average precipitation over a specified time period) is usually established somewhat arbitrarily. Daily precipitation is compared to evapotranspiration to determine the rate of soil-moisture depletion, and express these relationships in terms of drought effects on plant behaviour. Operational definitions, however, require weather data on hourly, daily, monthly, or other time scales and, possibly, impact data (*e.g.*, crop yield) to analyse drought frequency, severity, and duration for a given historical period.

Probability of occurrence of drought in different meteorological sub-divisions of India

Meteorological sub-division	Frequency of deficient rainfall (75% of normal or less)
Assam	Very rare, once in 15 years
West Bengal, Madhya Pradesh, Konkan, Bihar and Orissa	Once in 5 years
South interior Karnataka, Eastern Uttar Pradesh and Vidarbha	Once in 4 years
Gujarat, Eastern Rajasthan, Western Uttar Pradesh	Once in 3 years
Tamil Nadu, Telangana, Jammu and Kashmir	Once in 2.5 years
Western Rajasthan	Once in 2 years

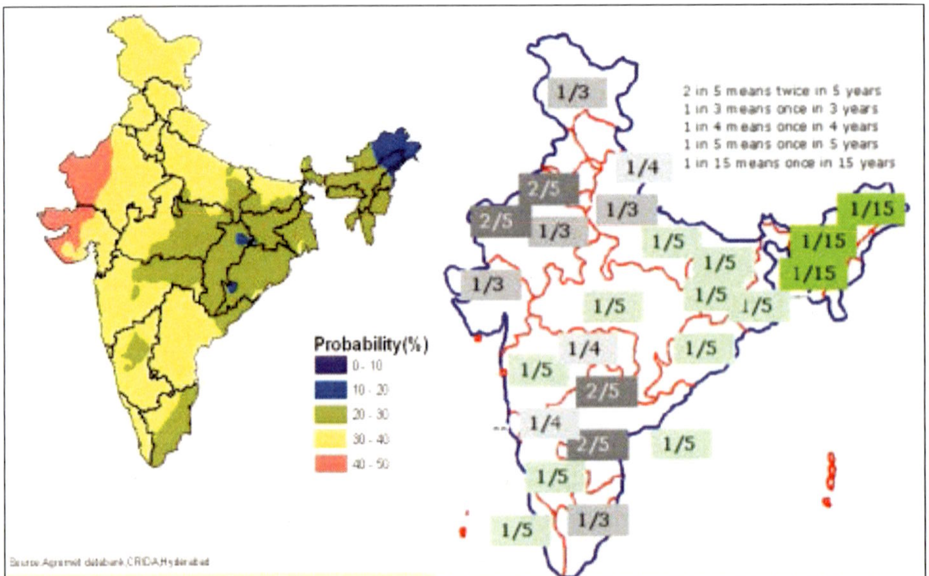

Probability of drought occurrence

- **Drought prone area:** Probability of a drought year is greater than 20%.
- **A chronic drought prone area:** Probability of drought year is greater than 40%.
- **When drought affected area in the country is 26–50% of the country:** Moderate drought.
- **>50% of the area in the country is under moderate or severe drought:** Severely affected by drought.

Classification of drought: Droughts can be classified into 3 categories based on general concept (a) Permanent drought (b) Seasonal drought and (c) Intermittent droughts. These categories are of regional classification based on the rainfall distribution in relation to potential evapotranspiration demand of the atmosphere. However, the droughts can be classified as follows in a given region.

I. Meteorological drought: *This is the situation when the quantum of annual rainfall decreases below a given normal (average) value say 25 per cent or less.*

Variety of meteorological definitions in different countries illustrates why it is not possible to apply a definition of drought developed in one part of the world to another. For instance, the following definitions of drought have been reported:

- **United States (1942):** Less than 2.5 mm of rainfall in 48 hours.
- **Great Britain (1936):** Daily precipitation less than 0.25 mm for fifteen consecutive days.
- **Libya (1964):** Less than 180 mm of annual rainfall.
- **Bali (1964):** A rainless period of six days.

Data sets required to assess meteorological drought are daily rainfall information, temperature, humidity, wind velocity and pressure and evaporation.

II. Hydrological drought: Hydrological droughts are usually related to meteorological droughts and their recurrence interval varies accordingly. Changes in land use and land degradation can affect the magnitude and frequency of hydrological droughts. Hydrological drought refers to a persistently low discharge and/or volume of water in streams and reservoirs, lasting months or years. Hydrological drought is a natural phenomenon, but it may be exacerbated by human activities.

Surface-water area and volume, surface runoff, stream flow measurements, infiltration, water-table fluctuations, and aquifer parameters are the data sets required to assess hydrological drought.

III. Agricultural drought: Agricultural drought focuses on precipitation shortages, differences between actual and potential evapotranspiration, soil-water deficits, reduced groundwater or reservoir levels, and so on. Plant water demand depends on prevailing weather conditions, biological characteristics of the specific plant, its stage of growth, and the physical and biological properties of the soil. A more clear definition of agricultural drought should account for the susceptibility of crops during different stages of crop development.

Deficient top soil moisture at planting may hinder germination, leading to low plant populations per hectare and a reduced yield.

Soil texture, fertility and soil moisture, crop type and area, crop water requirement, pests and climate are the data sets required to assess agricultural drought.

Types of agricultural drought

Early season drought: Early season drought occurs when there is delay in the onset of rains during the cropping season. Sometimes early rains occur tempting the farmer to sow the crops followed by a long dry spell leading to withering of seedling and poor crop establishment.

Mid season drought: Mid season drought occurs due to dryspell during the crop growth phases. Drought condition during vegetative crop growth phase might result in stunted growth, reduced leaf area development and even reduced plant population.

Late season drought: due to early cessation of rains, some time crops may encounter moisture stress during its reproductive phase which leads to forced maturity of the crops due to rise in temperature.

IV. Physiological drought

It occurs due to some physiological aberrations often plant roots fail to absorb water and nutrients from soil due to increased concentration of soil solution leading to this kind of drought. Presence of soluble salts in soil water increases the concentration of the solution and hence availability of water to plants is reduced, although sufficient water remains in the soil. Similarly presence of certain chemicals in the soil interferes with the water absorption by plants *e.g.*, humic acid restricts the water absorption in bog soils and the term is used as physiological dryness of the soil: however, plenty of water remains in the soil.

Crop adaptations to drought

The ability of plant to grow under water stress condition is called as drought adaptation. Plants survive under moisture stress conditions by two ways: (1) Escaping drought and (2) Drought resistance

I. Escaping Drought

Escaping drought is a classical adaptive mechanism which involves rapid plant development to enable the completion of the full life-cycle prior to a coming drought event. This strategy is widely used in populations of native plants, and is also applicable to Short duration cereal and pulse crops *viz.*, millets, cowpea, greengram, blackgram.

II. Drought Resistance

Drought resistance is defined as the mechanism causing minimum loss of yield in a water deficit environment relative to the maximum yield in a water constraint free management of the crop. Plants have evolved several mechanisms to cope with water deficit stress which includes drought avoidance and drought tolerance.

(i) Drought avoidance (Avoiding Stress)

'Drought avoidance' is the ability of plants to maintain (relatively) higher tissue water content despite reduced water content in the soil. This is achieved through a variety of adaptive traits involving the minimization of water loss (water savers) and optimization of water uptake (water spenders). Water spenders achieve higher tissue water status by maintaining the water uptake through increased rooting, hydraulic conductance *etc.*, under drought stress. In contrast, water savers use water effectively through reduced loss of water by reducing transpiration, transpiration area, radiation absorption *etc.*, under drought stress. Water savers avoid the drought by closing of stomata, lipid deposition on leaves, reduction in leaf area, deposition of wax on surface of leaves, spines and thick cuticle. Water spenders avoid the drought by well branched deep root system, rapidly growing root system, high root to shoot ratio and by adjusting the osmotic potential.

(ii) Drought Tolerance

'Drought tolerance' is the ability of plants to endure low tissue water content through adaptive traits. These adaptive traits involve maintenance of cell turgor through osmotic adjustment and cellular elasticity, and increasing protoplasmic resistance.

V. Socio–economic drought: It is associated with the supply and demand of some economic goods with elements of meteorological, hydrological and agricultural drought. Its occurrence depends on the supply of many economic goods, such as water, forage, food grains, fish and hydroelectric power, which depend on the weather. Due to the natural variability of climate, water supply is ample in some years, but insufficient to meet human and environmental needs in other years.

Human and animal population, growth rate, water and fodder requirements, severity of crop failure, industry type and water requirements are the data sets required to assess socio-economic drought.

Meteorological drought indices

The simplest way of classification of meteorological drought is deviation of annual rainfall quantum from the normal values.

Per cent of Normal Precipitation (PNP): The per cent of normal precipitation is the ratio of actual to normal precipitation for a given location and a given period, expressed as a percentage. Analyses using the per cent of normal are effective when used for a single region and a single season. In other applications, the index can vary depending on the choice of period, including monthly, seasonal or annual.

Standardized Precipitation Index (SPI): Certain deficit of precipitation has different impacts on the soil moisture, ground water, reservoir storage, snowpack and stream flow. The SPI was designed to quantify the precipitation deficit for multiple time scales. These time scales reflect the impact of drought on the availability of the various types of water resources. Soil moisture conditions respond to precipitation anomalies on a relatively short scale, while ground water, stream flow and reservoir storage reflect long-term precipitation anomalies. For these reasons, the SPI was originally calculated for 3, 6, 12, 24 and 48 month time periods.

The SPI is an index based on the precipitation record for a location and chosen period (monthly or years). The record is fitted to a probability distribution which is then transformed into a normal distribution so that the mean SPI for the location and period is zero. The index is negative for the drought and positive for non-drought conditions.

SPI classification	
2.0 or more	Extremely wet
1.5 to 1.99	Very wet
1.0 to 1.49	Moderately wet
–0.99 to 0.99	Near normal
–1.0 to –1.49	Moderately dry
–1.5 to –1.99	Severely dry
–2.0 and less	Extremely dry

Palmer Drought Severity Index (PDSI): In 1967, Palmer developed an index to measure the departure of the moisture supply. The PDSI is a meteorological drought index that is responsive to abnormal weather conditions, either on the dry or wet side. Palmer based his index on the supply-and-demand concept of the water-balance equation, taking into account the precipitation deficit. The objective of the Palmer Drought Severity Index (PDSI) is to provide standardized measurements of moisture conditions, so that comparisons can be made between locations and between durations. The Palmar drought classification is very popular in United states, Canada and some European countries.

PASI classification	
4.0 or more	Extremely wet
3.0–3.99	Very wet
2.0–2.99	Moderately wet
1.0–1.99	Slightly wet
0.5–0.99	Incipient wet spell
0.49 to –0.49	Near normal
–0.5 to –0.99	Incipient dry spell
–1.0 to –1.99	Mild drought
–2.0 to –2.99	Moderate drought
–3.0 to –3.99	Severe drought
–4.0 or less	Extreme drought

Water balance approach: After the popularization of Thornthwaite's (1948 and 1955) climate water balance, a strong school of thought came into existence on drought climatology on water balance approach. With the help of annual distribution of rainfall and potential evapotranspiration, climate water balance can be computed using Thronthwaite's book keeping procedure. Using the water balance parameters, Thronthwaite proposed an index known as **"Aridity index"**, which is a percentage ratio of annual water deficit to annual water need (or potential evapotranspiration).

$$\text{Aridity Index} = \frac{\text{Water Deficit}}{\text{Potential Evapotranspiration}} \times 100$$

Using this aridity index over a time-series, Subrahmnyam and Subramanian (1964) evolved a drought classification based on the deviation of aridity index, over its mean value with intervals of its standard deviation.

Reclamation Drought Index: It is a tool for defining severity and duration, and for predicting the beginning and end of a drought period.

RDI classification	
4.0 or more	Extremely wet
1.5–4.0	Moderately wet
0–1.5	Normal to Mild wetness
0 to –1.5	Normal to Mild drought
–1.5 to –4.0	Moderately drought
–4.0 and less	Extremely drought

Agricultural drought indices: In India, several workers have assessed the agricultural drought on the basis of water availability periods and aridity index.

Crop Specific Classification: Looking into the problems of drought classification, Sastri *et al.* (1981) proposed a new scheme of drought classification in which the concept of minimum requirement of index of moisture adequacy, IMA, a percentage ratio of actual evapotranspiration to potential evapotranspiration, AE/PE for obtaining average productivity of particular region was adopted. For obtaining average productivity, a minimum value of moisture status is required.

Crop Moisture Index (CMI): Palmar (1968) proposed drought classification based on Crop Moisture Index (CMI), which is a measurement of cumulative transpiration deficit. The Crop Moisture Index (CMI) uses a meteorological approach to monitor week to week crop conditions. It was developed by Palmar (1968) from procedure within the calculations of the PDSI. The PDSI monitors long term wet and dryspells, where as the CMI was designed to evaluate short term moisture conditions. It is based on mean temperature and total rainfall for each week, as well as the CMI value from previous week. The CMI is not good for long term drought monitoring tool. It is also not applicable during seed germination at the beginning of a specific crop's growing season.

Crop and moisture distribution specific classification: Though the above classification is more rational and crop and productivity level specific, it was found that the intermittent dryspells which affect different crops differently at different growth stage were not considered.

To overcome this an integrated scheme of drought classification was developed (Ramana Rao *et al.*, 1981 and Sastri and Patel, 1984). In this, drought intensity at different growth stage like seedling, vegetative, reproductive and maturity stage was worked out to obtain the overall drought intensity during that particular crop growth season. Based on this, the productivity trend under different moisture stress levels were worked out to examine the effect of technology, which otherwise was not discernible due to large inter annual variability of the productivity (Sastri *et al.*, 1983).

Impacts of drought: There are several impacts which are commonly referred to as direct and indirect. Direct impacts include reduced crop. rangeland, and forest productivity, increased fire hazard, reduced water levels, increased livestock and wildlife mortality rates, and damage to wildlife and fish habitat.

The impact of drought are commonly classified as economic, environmental and social (Wilhte, 1992).

1. Economic Impacts: In agriculture and related sectors, economic impacts occur due to reliance of these sectors on surface and ground water supplies. In addition to losses in yields in crop and livestock production, drought is associated with insect infestations, plant disease and wind erosion. The incidence of forest and range fires increases substantially during extended periods of droughts, which in turn places both human and wildlife population at higher level of risk.

Income loss is another indicator used in assessing the impacts of drought. Retailers and others who provide goods and services to farmers face reduced business. This lead to unemployment, increased credit risk for financial institutions, capital shortfalls, and eventual loss of tax revenue for local, state, and federal governments. Prices for food, energy, and other products increases as supplies are reduced. In some cases, local shortages of certain goods result in importing these goods from outside the drought- stricken region. Reduced water supply impairs the navigability of rivers and result in increased transportation cost because products must be transported by alternative means. Hydropower production may also be significantly affected.

2. Environmental Impacts: Damages to plant and animal species, wildlife habitat, and air and water quality, forest and range fires, degradation of landscape quality, loss of biodiversity, and soil erosion are the environmental losses due to drought. Some of these effects are short-term, conditions returning to normal following the end of the drought. Other environmental effects last for some time and may even become permanent. Wildlife habitat, for example, may be degraded through the loss of wetlands, lakes, and vegetation. However, many species eventually recover from this temporary aberration. The degradation of landscape quality, including increased soil erosion, may lead to a more permanent loss of biological productivity.

3. Social impact: It includes public safety, health, conflicts between water users, reduced quality of life, and inequities in the distribution of inputs and disaster relief. Many of the impacts identified as economic and environmental have social components as well. Migration of people and animal is a significant problem in many countries due to drought, often stimulated by a greater supply of food and water elsewhere. Migration is usually to urban areas within the stressed area, or to region outside the drought area. Migration may even be to adjacent countries. When the drought has abated, the migrants seldom return home, depriving rural areas of valuable human resources. The drought migrants place increasing pressure on the social infrastructure of the urban areas, leading to increased poverty and social unrest.

Drought management

Drought is indeed a many faceted natural disaster that leads to serious socio-economic impacts, which have long (Sastry, 2003) term implications. It is in this context that development of appropriate drought management strategy is of great importance.

At present, the main objectives of drought management strategy are:

- Distribution of essential commodities such as water, fodder and food at subsidized rates.
- Optimum utilization of resources in the affected areas with emphasis on primary resource *viz.* soil, water, vegetation, man power *etc.,* relief works undertaken for providing employment to drought stricken population mainly for drought proofing.
- Improvement in terms of living conditions of the rural poor who suffer the most due to scarcity and drought in particular, and the community in general by creating direct and indirect employment and taking up short duration programmes of development.

Drought management is currently addressed by the following mechanism and sectors

Governmental

1. Policy issues, national, regional, and district level,
2. Rural development infrastructure
3. Input supply, marketing and farm advisory services

Non-governmental

1. NGOs (Non-Governmental organization)
2. Rural institution and local self governments
3. Private sectors
4. Philanthropic organizations community codes (tribes, herders)
5. International aid agencies.

Research and development institutions

1. Weather forecasts,
2. Best practices for rainwater and soil management through linking on station and on farm research,
3. Contingency crop planning/mid season correction,
4. Alternative land use system
5. Use of remote sensing and geographical information system

Drought mitigation

Drought is a recurring phenomenon and its occurrence cannot be avoided. However, its impact can be minimized through application of science and technology in developing suitable drought management plans. Mitigation action specific to drought can be defined as short and long term actions programmes or policies implemented in advance of drought that reduce the degree of risk to people, property and increases productive capacity (Das, 1999).

Some of the drought mitigation strategies help in effective response and mitigate the hardship of the people as follows:

- Arrangement of reasonable buffer stock of food grain and fodder.
- Ensure supply of good drinking water in rural areas for human and livestock in drought affected areas.
- Assess fodder requirement in drought affected district and locate areas where shortage are likely to occur and arrange supplies from outside.
- Fodder cultivation to be encouraged wherever feasible.
- Rejuvenation of traditional rain water systems *viz*; rivers, tanks, nalas *etc.*
- Rain water harvesting for both drinking and cropping.
- Management of human, livestock population to reduce pressure on fragile arid ecosystem.
- Timely availability of credit, postponement of revenue collection and repayment of short term agricultural loans.
- Appropriate land use planning (intercropping system), discouraging water intensive crop and encouraging drip and sprinkler system.
- Creation of local task force in each district to initiate relief measures immediately after the drought takes place.
- Implementation of crop and allied livestock insurance schemes.
- Provisions for cattle camps in drought affected areas.
- Early warning and drought monitoring should be carried out on the basis of long, medium and short term forecast.

Management of agriculture drought

Late onset of monsoon, prolonged dryspells during monsoon season or early cessations of rain was the major characteristics of monsoon aberration. These situation are sought to be managed through a two prolonged strategy, viz., selection of suitable crops and cropping systems in relation to drought and conservation and efficient utilization of water.

Crop management in relation to drought

In rainfed regions, crop production depends on monsoon rainfall both in terms of quantity and its distribution. Since the rainfall is unpredictable, there is a need to adopt suitable crop management practices for stable productivity. Normally the aberrant weather situation that prevail in rainfed region are:

timely onset of monsoon followed by immediate dry spell, delayed onset of monsoon, early withdrawal of monsoon and extended monsoons.

I. Timely onset of monsoon followed by immediate dry spell

During some years, a dry spell may occur immediately after sowing of crop. This results in poor germination due to soil crusting. There may be too many gaps either in a row or in rows. This situation can be broken by gap filling with same crop. Alternatively, transplanting of excess seedlings from an area can be considered. In crops like Fingermillet, chilli, pigeonpea, maize, cotton, *etc.*, emergency nurseries can be raised in the backyard or wherever the water facility is existing so as to use the seedlings for gap filling or transplanting in the fresh area during exigencies. Under extreme situations of very poor germination, if sowing window is still open for that crop a medium or short duration variety can be resown in the affected area *ex:* In fingermillet, if july sown crop is not germinated due to dryspell, a medium duration variety *viz;* GPU-28, ML-365, Indaf-5 can be sown in the first fortnight of august. By chance if there is reduced germination of these medium duration varieties due to drought, since sowing window for fingermillet is still prevailing short duration variety *viz*; GPU-48 can replace these medium duration varieties for sowing till end of august or first week of September.

II. Management of intermittent dryspells (drought): if the drought occurs during vegetative stage of crop growth, it may result in stunted growth, low leaf area development and even reduced plant population. Suitable management practices for mitigating dryspell at early and mid season crop growth stages are as follows.

1. Mulching: Mulching helps in conserving soil and moisture. It is a protective covering placed around the plants or in between rows and mulch act as mechanical barrier for preventing evaporation of soil moisture and weed growth, the freezing of roots and the growth of weeds.

Types of Mulches

(a) Organic mulches (b) Inorganic mulches

Organic mulches: It is the use of living material and as such will decompose over time. During their decomposition vital nutrients will be added back into the soil.

Compost: It mulches and feeds the soil as it decomposes. This mulch is applied at the depth of 1–3 inches.

Pine needles: It is commonly used in acid soils. Cheap and allow water to pass through freely to the soil below. It decomposes slowly. It should be applied at the depth of 1–1.5 inches.

Straw: It provides great insulation, water penetration and weed control. Care should be taken that the straw does not contain weed seeds itself. It should be applied at a depth of 6–8 inches.

Grass clippings: It is readily available and decomposes quite quickly adding nitrogen back into the soil. It should not be added too fresh as it can heat up quite a bit and possibly cause damage to plants. It should be applied at a depth of one inch.

News paper: It provides great weed control and is readily available. Apply mulch on the top to keep it in place. Apply in two layer sections.

Inorganic mulches: These are inert materials that have not originated from living material.

Stone: It provides great insulation. It degrades very slowly. Apply at 2–4 inches depth.

Plastic: It does not decompose, so it does not add anything into the soil. It acts as a great weed control and is easily laid. It must be perforated to allow water to pass through. It should be applied in a thickness of 1–6 mm.

Geo textile: Expensive blanket like synthetic fiber provides great weed control and allows for water penetration. It is commonly used in conjunction with cover mulch (ex: bark chips). It should be applied in a single layer.

Thickness of mulch layer: Thickness of the mulch is very important factor to be considered before use of mulch in crop production. If mulch is too thick (depth >3–4 inches) poses secondary problems. Suppose fine organic mulches such as compost (or) shredded leaves are spread 2–3 inch layer where as coarse material such as woodchips need to spread 3–4 inch layer.

Advantages of mulching

1. **Prevents soil erosion:** Soil erosion happens when winds or water slowly wear away the surface of soil and remove it. Mulching the soil will prevent this as there is a protective replaceable layer that comes between the surface of soil and forces of nature above it.
2. **Insulates the soil in winter:** The soil in winter will insulate the soil preventing it from repeated patterns of freezing and thawing and will insulate plant roots. It prevents heaving of soil in spring.

3. **Cools down the soil in summer:** A layer of mulch in summer protects the soil beneath from the extreme heats of summer and decrease the need for constant watering *i.e.*, water retention is improved.
4. **Helps against soil compaction:** The layer of mulch acts as a buffer (or) extra layer between entities that can cause compaction on the soil below equipment, heavy rains, *etc*.
5. **Prevents weed growth:** One of the more useful benefits of mulching is weed prevention. While the mulch itself will not stop weeds from germinating, it serves to act as a barrier between the weeds and the outside world above, when a weed reaches the mulch layer it will not be able to break above it will eventually dieback. Some mulch performs this weed prevention feature better than others.
6. **The greatest benefit from plastic mulch** is raising the soil temperature in the planting bed to promote faster crop development and earlier yields. Black plastic mulch can result in an earlier harvest.
7. **Soil water loss is reduced under plastic mulch.** As a result, more uniform soil moisture is maintained and irrigation frequency may be reduced.
8. **Reduced fertilizer leaching:** Excess water runoff the impervious mulch. Fertilizer beneath the mulch is not lost to leaching. So fertilizers are efficiently used.
9. **Cleaner product:** The edible product of mulched crop is clean and not subjected to rots because soil is not splashed on the plants (or) fruits.
10. **It aids in fumigation:** Mulches increase the effectiveness of soil fumigant chemicals because of the impervious nature of the plastic mulch; it acts as a barrier to gas escape.

Disadvantages of Mulching

1. **Removal of disposal:** A major problem with plastic mulch is removal from the field after cropping. Plastic mulch especially black plastic, does not breakdown and should never be disked into the soil.
2. **Need for specialized equipment:** The use of plastic mulches require specialized equipment including a bed press, mulch layer and mulch transplanter and plug mix seeder. These investments are more difficult to justify on small acreages, unless farmers co-operatively purchase the equipment.
3. It is a labour intensive.
4. Fire hazards.
5. Termite attack in dry condition.

Examples of mulches and their characteristics

Mulch	Advantages	Disadvantages	Reapplication time
Organic peat	1. Coarse textured 2. Increase WHC of soil 3. Over time cause increase in acidity	1. Expense 2. Difficult to remoisten after it has dried 3. Repels water.	1 year
Wood chips/ shredded wood	1. Decompose slowly 2. Improve the soil 3. Long lasting and attractive. 4. Relatively inexpensive. 5. Easily adoptable	1. Injury to plants may occur if applied too heavy 2. Wood chips may not cling well on steep slopes 3. Do not use near building foundation.	1–2 year depending on size and type of wood.
Straw	1. Inexpensive. 2. Readily available.	Coarse appearance may contain weed seeds and some diseases	Needs frequent application.
Lawn clippings	1. Readily available. 2. Usually free. 3. Decompose slowly. 4. Improve the soil.	1. May mat down and interfere with water and air movement, if applied too thickly. 2. May contain weed seeds. 3. May contain herbicides(if herbicides used to control weeds).	One season.
Leaves	1. Readily available. 2. Usually free. 3.Decompose slowly 4. Improve the soil.	1. May mat down and interfere with water and air movement if applied too thickly. 2. Should be shredded and partially decomposed before using.	One season.
Pine needles	1. Fragrant. 2. Decompose slowly. 3. Improve the soil. 4. Overtime will increase the soil acidity.	1. Not always available.	1–2 years.
Cocoa bean hulls	1. Attractive and fragment. 2. Good soil conditioner.	1. Tends to float away during heavy rains. 2. Expensive. 3. Will decompose by mildewing.	One season.

Mulch	Advantages	Disadvantages	Reapplication time
Inorganic plastic	1. Completely inhibits weed growth.	1. Crack easily from low temperature or UV instantly allowing weed growth. 2. Expensive. 3. Allows no water or air to enter. 4. Stressing roots under mulched plants needs a supplemental irrigation	Variable.
News paper	1. Prevents weed germination. 2. Effective between vegetable rows. 3. Readily available.	1. Can blow away if not weighed down. 2. Un sighting in landscape setting.	One season.
Land scape fabric	1. Allows water and permeates. 2. Desirable. 3. Suppresses most weeds.	1. Weeds may germinate on top of fabric. 2. May shed mulch cover on steep slopes. 3. Cover with light top mulch to present fabric deterioration.	Indefinite

Mulching is an important agronomic measure that not only dissipates the kinetic energy of raindrops but also prevents soil erosion and facilitates infiltration and reduces runoff and evaporation losses. Frequent intercultivation between crop rows create dust mulch which is an useful operation helps in breaking the soil crust. It breaks capillary conductivity and minimises evaporative loss of water. If break in monsoon is brief, dust/soil mulch is good tool for extended survival of crop plants.

Crop residues such as sorghum and maize stubbles, dry grass, wheat straw and pigeonpea stalk can be used as surface mulch. These mulches prevent moisture loss and prolong the moisture retention period. Studies in black soils at Solapur over a period of 3 years have indicated that increased sorghum yield by 50–70%.

Mulch upon decomposition increases the organic matter content of the soil, helps in higher infiltration and moisture retention. Studies in red soils at Bangalore have indicated that incorporation of maize residue @ 4 t ha^{-1} increased moisture retention. The crop yields on such plots are 80% higher than that of control during the following year (Anon., 1984).

Additional intercultivation during dryspell: At vegetative stage (early drought) enhanced the productivity of castor, sunflower and pigeonpea by 15–20% compared to no interculture (Subbareddy *et al.*, 1996).

2. Nutrient management

- Band placement of fertilizer preferably by seed-cum-fertilizer drill.
- Split application of N.
- Balanced fertilization to suit the crop requirement.
- Organics like FYM, compost, crop residue, green manures should be applied.
- Where ever possible, bio-fertilizers to be used (PSB).
- *In situ* moisture conservation to improve nutrient use efficiency.

Additional N application: After the relief of drought, additional N application should be made through top dressing. In light textured soils, top dressing of N after the release of stress with the receipt of rains will give immediate boost to the suffered plants for want of nitrogen.

3. Altering plant population : If the moisture stress occurs at early stage, thinning of plant population needs to be considered. Thinning of plant population by removing every third row at the time of drought increased the productivity of sorghum from 1550 to 2110 kg ha^{-1} at Bellary in Karnataka (Venkateswaralu *et al.,* 1991). Under extreme situations, removing every alternate row can reduce the evapo-transpirational demand and save the crop. The removed plants may be used as fodder or as mulch *in-situ.*

In drought prone areas, providing wider spacing (Table 68) will not only reduce the plant population but also facilitate frequent inter cultivation for weed control, soil mulch and hence conservation of moisture and drought escape.

Table 68: Spacing recommended for different crops:

Crops	Spacing
Pearl millet, rabi sorghum, safflower, castor, cotton	60 cm
Horsegram, sesamum, green, ground nut	45 cm
Redgram	90–120 cm
Pearl millet and sunflower	60–135 cm

4. Antitranspirants: Plants transpire water vapours continuously from all above ground parts particularly through leaves. This process of evaporation of water from the aerial parts of plants is termed as transpiration. Approximately 99 per cent of the water taken by the plants roots is transpired to the atmosphere. Transpiration occurs through different types of apertures such as cuticles, lenticels and stomata. Among these apertures stomata accounts for 90–97 per cent of transpiration. Transpiration is considered as a necessary evil. Reduction in transpiration may help in maintaining favorable water balance in dry farming. Any material that is applied to plant surfaces with the aim of reducing or inhibiting water loss from plant surface is called antitranspirants.

Use of antitranspirants reduces the rate of transpiration and brings economy in water utility by crop plants.

Scope of antitranspirants

- Under dryland area to reduce water losses through transpiration.
- In costly irrigation facilities for extending the irrigation intervals.
- For reducing transplanting shock of nursery plants

Types of antitranspirants: Based on mode of action, antitranspirants are of four types

Stomata closing type: Most of the transpiration occurs through stomata on the leaf surface. Spray material used for various purposes such as certain fungicide PMA, 2,4–D and Atrazine at low concentration and metabolic inhibitors have been found to cause the closure of stomata and thereby reduction in transpiration. The effectiveness of an antitranspirant is also depends on the coverage of lower surface of leaves, interaction of the antitranspirant with external environments, surface anatomy of leaves and rate of formation and growth of new leaves.

Film forming type: Foliar spray of waxy or plastic emulsions such as mobileaf, hexadecanol and silicon produce an external physical barrier outside the stomatal opening to retard the escape of water vapour through stomatal opening. The film so formed should have more resistance to the passage of water than to that of carbon dioxide. Film type antitranspirants which provide selective type of permeability barriers to water vapours and carbon dioxide diffusion in the required directions have not yet been found so far.

Reflectance Type: White reflecting materials such as Kaolin, china clay and lime water spray form a coating on the leaves and increase the leaf reflectance (albedo). By reflecting the large amount of radiation, they reduce leaf temperature and vapour pressure gradient from leaf to atmosphere and thus reduce transpiration. Application of 5% kaolin spray has been found to reduce transpiration losses markedly.

Growth retardants: Foliar application of chemicals such as Abscissic acid, cycocel reduces shoot growth, increases root growth and induces stomatal closer. Thus, the application of such chemical helps in improving the water status in the plants and soil.

Limitations of antitranspirants

- May reduce the rate of photosynthesis
- May increase the leaf temperature by reducing evaporative cooling
- Interaction of climatic factors with antitranspirants reduces their effectiveness for longer duration
- Sometimes marginal cost is more than marginal returns
- May produce toxic effects on leaves

5. Stripping of older leaves : Striping of bottom most older and diseased leaves can reduce substantially evapo-transpirational demand of the crop and helps to direct moisture and nutrients to actively growing part of the plant. In sorghum, normally 10 to 15 leaves are produced. When plants are at flag leaf stage, stripping of bottom 3–4 leaves helps in drought mitigation. This practice may save moisture to certain extent and stripped leaves can be used as fodder or as mulch.

6. Ratooning: In *kharif* season where intermittent dryspell range 4–6 weeks, crop productivity is adversely affected. Under such conditions, ratooning practice can be considered especially in sorghum and pearl millet. In pearl millet when dry spell exceeds 30–35 days, harvest the crop for fodder and allow ratoon for further growth with subsequent rainfall. Similarly in crops like maize, instead of allowing the crop for grain purpose, it can be harvested for green cobs and fodder if the crop is suffering from deficit moisture. While in field bean and pigeonpea, crops can also be harvested for green pods for vegetable purpose. For quick rejuvenation, supplementation of additional N fertilizer is suggested. This practice can be extended to *kharif* S*orghum* areas under similar conditions. If there is a complete failure of rains and crop is not in a position to produce any seed yield in *kharif* season, in double cropped areas (sorghum, groundnut, *etc.,*) it may be harvested for fodder and the land could be kept ready to make use of showers for *rabi* season cropping.

III. Delayed onset of monsoon

1. Seed hardening: Pre-sowing seed hardening is one of the methods of modifying the physiological and biochemical nature of seed to get favourable characters for drought tolerance. Subjecting them for several days to moderate dehydration, low temperature or treatments with growth retarding substance can increase drought tolerance. The beneficial effects of seed hardening include accelerated rapid germination and growth rate of seedling, quick recovery from wilting, resistance for drought, early vigour and high yields. Seed hardening involves seed soaking in solution or water for a specific period (16 hrs) at room temperature followed by shade drying (24 hrs).

Physiological basis for seed hardening

(a) It ensures hydration of colloids
(b) It increases viscosity and elasticity of protoplasm
(c) It increases the bound water content
(d) Hardening of seeds increases the intensity of photosynthesis
(e) It increases the hydrophilic colloids and decreases lyophilic colloids
(f) Seed hardening increases the temperature requirement for protein coagulation
(g) It intensifies respiration
(h) It increases the water balance of the plant
(i) Seed hardening develops more efficient root system

Biochemical changes induced by priming

The positive effects of priming on the germination performance of many species are attributed to the induction of biochemical mechanisms of cell repair: the resumption of metabolic activity can restore cellular integrity, through the synthesis of nucleic acids (DNA and RNA), proteins and the improvement of the antioxidant defence system.

(a) In spite of its crucial role, the amount of DNA which is needed in the repair processes is only 20–30% of the total DNA synthesized during priming. The rest is mainly represented by mitochondrial DNA; in fact, the number of mitochondria was shown to rapidly increase during priming in leek seeds(*Allium porrum*).
(b) Priming percent has no direct effect on cell division, but advances its beginning (G1 and G2 phase of mitosis) from phase III to phase II of seed imbibition. This advance is enabled by an accumulation of β-tubulins in primed seed, which are proteins involved in maintaining the cell cytoskeleton and forming the microtubules necessary to cell division
(c) rRNA is as much necessary to repair cell damages as DNA. Priming allows the recovery of rRNA integrity, in turn ensuring a correct coding of amino acids for the synthesis of proteins during seed germination.
(d) In protein synthesis, a compensation mechanism has been elucidated. Protein synthesis, an essential requisite for germination, starts several minutes after hydration. At this stage, osmopriming depresses protein synthesis in the embryo and reserve tissues, compared to seed soaked in water for the same time. Then, priming induces a higher level in the subsequent germination.

(f) Osmopriming induces the synthesis and activation of enzymes catalyzing the breakdown and mobilization of reserve substances. This effect is associated with the water deficit induced by osmopriming, which is supposed to determine a mobilization of reserve proteins.

Beneficial effect of seed hardening

(a) Acceleration of germination, growth rate of seedlings and flowering
(b) Favours quick recovery from wilting
(c) Resistance or tolerance to salinity and drought
(d) Ability to withstand higher temperature for prolonged periods (up to two days) without loss of viability
(e) Quick germination of seedlings favours more effective competence with weeds
(f) Increase in growth and yield of crop plants

Materials and chemicals used for seed hardening

(a) Water
(b) Aqueous salt solutions like chlorides of sodium, calcium and potassium sulphates of sodium and ammonium, potassium nitrate *etc.*
(c) Growth regulators like Gibberellic acid, Chloro choline chloride, Kinetin, 2 Chloro ethyl phosphonic acid, Ascorbic acid *etc.*
(d) Vitamins like K3, Nicotinic acid, Pantothanic acid *etc.*
(e) Plant products like Garlic extract, Coconut water *etc.*
(f) Osmotic priming materials like D-Mannitol. Polyethylene glycol *etc.*

2. Selection of appropriate crops and crop varieties: deficit in stored soil moisture stress lead to drought which in turn lead to affect crop production adversely. The strategy to overcome deficient stored soil moisture in dryland is to identify suitable crops that will require less moisture to complete its life cycle. Another strategy is to identify suitable varieties based on rainfall.

Soil moisture holding capacity depends on soil structure. The moisture holding capacity of sandy soil is less than clay or silt loam soils. Hence, during the period of low or deficit rainfall, the crop growing on sandy soils is more prone to drought than to crops growing in clay soils. Based on the moisture holding capacity of soil, long duration rice variety is recommended for fine textured and short duration variety for coarse textured soils.

Some crops have the ability to withstand drought because of extensive root system. Sorghum is more tolerant to drought than maize because of different root concentration. Greater drought resistance is shown in plants with deeper and more extensively branched root system.

3. Dry seeding: whenever onset of rains get delayed, it is preferable to go for dry seeding with necessary seed treatment if seed bed is timely prepared. This practices helps to avoid further delay in sowing that could happen once the showers of high intensity are received in medium and deep black soils. Dry seeding has to be followed based on the probability (> 60%) of rains.

Sowing of Finger millet and bolded seeded crops *viz;* castor, pigeonpea is found to be advantageous with this type of technology. Soil should be necessarily dry for sowing otherwise seeds get damaged under semi-wet condition leading to poor plant stand establishment.

4. Raising nursery and transplanting: In endemic areas, community nurseries using farm pond water or tank water need to be raised and seedlings could be transplanted with receipt of rains. In Bangalore, medium and late varieties of finger millet with transplanting maintained yield levels under delayed condition. Transplanting of finger millet gave 25% higher yields compared to direct seeding in Finger millet-pigeonpea (8:2) intercropping system in rainfed Alfisols (Subbareddy and Havanagi, 1992). The limitation of this technology is the availability of water for raising nurseries and availability of labour on the day of receipt of rainfall for transplanting. In crops like chilli, pigeonpea, maize, finger millet, wherever the water facilities are existing nurseries can be raised for transplanting under field conditions upon receipt of rains.

5. Wider row spacing: Wider rows facilitate inter cultivation and earthing up creating micro relief for *in-situ* rain water harvesting. This is more advantageous in maize, jowar, cowpea, pigeonpea, castor, sunflower, *etc.* When soil moisture is limited, it would be advantageous to increase inter row spacing and decreasing intra row spacing. Intra row competition would result in smaller plants and in inter row, reservoir of soil moisture would be maintained for longer period of growing season. Wider rows apart from giving the advantage of rapid land coverage, efficient weeding, prolonged inter cultivation and dust mulching can be done even in low rainfall years. Wider row spacing helps to develop more proliferous root systems, better water extraction in initial stages (Umbrani and Bhoi, 1984).

IV. Early withdrawal of monsoon:

If monsoon terminates earlier than the usual, it cuts the length of growing season. Normally crops are in their maturity stage and those growing on shallow soils suffer heavily. Sowing of *rabi* crops also gets affected.

1. Protective irrigation: Life saving irrigation or supplemental irrigation from rainwater collected in farm ponds is one of the important practices to mitigate drought experienced during the life cycle of crop. It is also called as deficit or upcoming irrigation, since it can be applied to save the crop in case of unexpected total seasonal rainfall.

In potentially double cropping areas, supplemental irrigation can be used for establishment of second crop and in the second crop, as the surface soil is dry enough to adversely affect the crop stand after *kharif* crop harvest. Among various methods of irrigations, alternate furrow irrigation was efficient for shallow rooted and narrow spaced crop like sorghum over furrow irrigation (Vijayalakshmi *et al.*, 1989). For a given depth of water, drip method of irrigation saved 50% of pond water compared to modified sprinkler and surface method of irrigation. Sub soil application of water was found advantageous in cutting down evaporation losses but difficult to implement in farmers fields.

2. Reduce the plant density: At Bangalore, it is suggested to harvest the crops like maize for fodder leaving intercrop for full maturity. In Northern region of Karnataka, it is suggested to thin *rabi* sorghum or safflower based on moisture availability within 40 to 45 days of sowing and in *rabi* sorghum + chickpea and safflower + chickpea system, uprooting sorghum and chickpea component, respectively was suggested.

V. Extended monsoon:

Such aberration is rarely experienced. This situation is favourable in medium to deep black soils where double cropping is feasible. After harvesting of *kharif* crop, attempt can be made to take short duration pulse or fodder crop (horse gram). In crops like *kharif* sorghum, ratooning can be practiced and additional grains could be obtained.

1. Pre mitigation measures for drought: The management practices recommended for soil and moisture conservation practices mentioned in chapter 6 can also be followed as pre mitigation measures.

2. Rainwater harvesting and watershed based development approach: There is a need to recharge aquifers and conserve rainwater through water harvesting structures. Apart from bridging the demand-supply gap, recharging improves the quality of ground water, raises the water table in wells and prevents flooding and choking of drains. One can also save energy to pump ground water as the water table rises. Based on this realization, the government initiated a massive programme for integrated watershed development in dryland areas. Further during the last two decades, other programmes were initiated by the government for mitigating the effects of drought. The ecological restoration like drought prone areas programme and desert development programme even though initiated earlier, has undergone considerable transformation in the same direction.

Farm ponds are very age old practice of harvesting runoff water in India. These are the bodies of water, either constructed by excavating a pit or by construction of embankment across a water course or the combination of both.

3. Timeliness: Sowing dryland crops with onset of monsoon rainfall can significantly improve the crop yields. Timely sowing helps in achieving optimum utilization of seasonal rainfall, reduce the incidence of pest and diseases. Lack of efficient implements and adequate draft power are the major constraints to timelines on farming operations.

4. Agroforestry: Growing of annual crops in arid regions is highly vulnerable to frequent risk posed by drought. Under such situation agroforestry provides multiple benefits in terms of providing food, fodder, fuel, fruits, fibers and high aromatic and industrial products and also provides security against risk.

5. Crop insurance: It was attempted for the first time in India during *kharif* 1985, when the comprehensive crop insurance scheme was launched. One of the basic objectives of the scheme was to provide a measure of financial support to the farmers in the event of crop failure following the drought effects. This was further modified and a modified comprehensive crop insurance scheme (MCCIS) was proposed from *kharif*, 1999.

6. Policy interventions: The government of India over the years comes up with a number of policies based on experiences gained from various drought situation and other impacts during the last few decades. The drought policies can be classified as: short term, medium term and long term policies.

7. Weather based agro advisories: A comprehensive drought management system would include the collection of climatological data (*e.g.*, temperature and precipitation) as well as stream flow, reservoir and groundwater levels, soil moisture, snow pack and remotely sensed data from satellites. This information is useful in forecasting of agricultural and meteorological drought. Monitoring and early warning techniques include the use of indices to track current drought situation and to review them in the historical context.

As drought is a phenomenon which would come up slowly in the absence of rainfall over the given region, Weather forecast in different temporal ranges helps early warning to the farmers to take appropriate measures to prepare, mitigate and combat drought. The agro meteorological advisory services helps them in this endeavour. Modern tools of meteorology have helped meteorologists to improve and extend their predictions and reduce the number of failures in weather predictions.

Information on impending weather 3–10 days in advance *i.e.*, medium range weather forecast is vital for effectiveness of farming practices like sowing of weather sensitive high yielding varieties, need based application of fertilizers, pesticides, insecticides, irrigation and harvest planning. To extend the forecast for 3–7 days, medium weather forecast for the benefit of farming community, the India Meteorological Department (IMD) has been issuing district level medium range forecast to the 127 agro meteorological field units in the country under the project integrated agro advisory services (IAAS). Presently, the IMD will provide district specific medium weather forecast valid for coming five days on Tuesday and Friday. Based on the forecast received from IMD, nodal officer or his associates will prepare the final forecast on quantitative rainfall, tendency in maximum and minimum temperatures, wind speed and direction, cloud cover and relative humidity valid for next 5 days.

In order to better serve the farming community, agro meteorological advisory services should focus on:

(a) Generating location specific weather forecasts in quantitative terms
(b) Develop decision tools that are calibrated against local data for use by farmers in combination with the observations of local weather
(c) Evaluate best tactical management decisions for farmers on the basis of local information and use of models and broadcast the information in real time together with the weather forecasts.

Contingency crop plans

In preparation of contingency plans not only long range forecast but also medium range weather forecast are taken into consideration and contingency plans are needed if following conditions prevail.

- Failure of south-west monsoon
- Delayed onset or early withdrawal of monsoon
- Deficit or erratic rainfall
- Damage to crops due to floods, cyclones *etc.*
- Crop loss due to drought
- Insufficient supply of irrigation water or late release of canal water
- Long dry spells
- Heat or cold waves
- Severe pest or disease outbreaks due to favourable weather conditions

The contingency strategies are to be based on location specific needs and situation based (rainfed or irrigated). Within the regions also vary with soil type. Monitoring of season during *kharif* and suggestions for contingent crops or strategies for Karnataka are as follows:

Contingent crop planning for varied weather conditions in red soils in Karnataka

Decision tree for contingency measures for drought

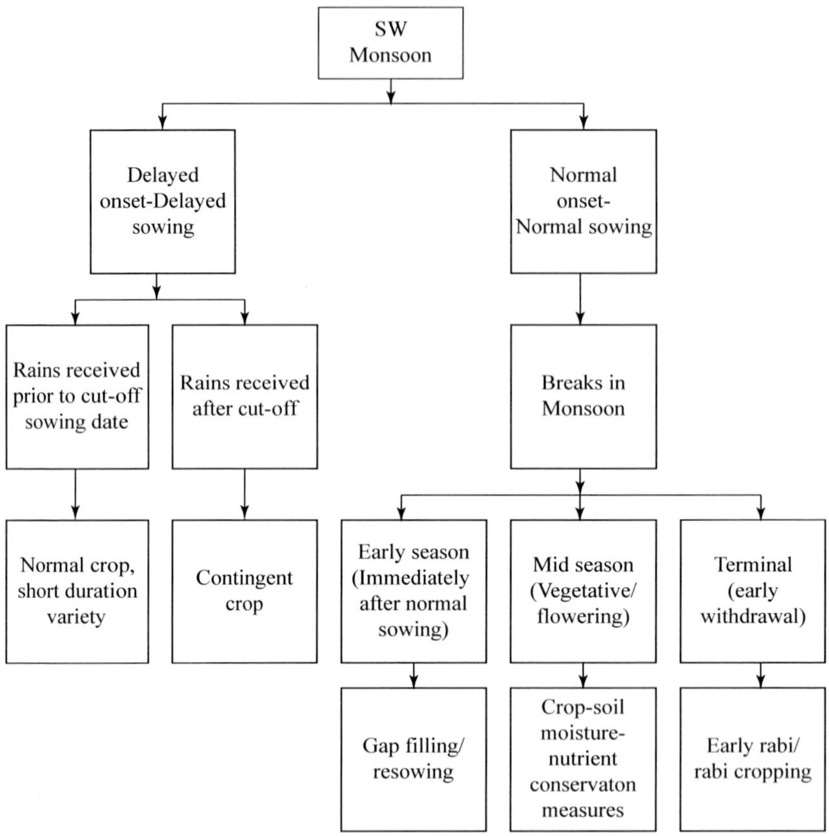

Contingent crop planning for varied weather conditions in black soil in Karnataka

1. Normal onset of monsoon is favourable for *kharif* crops

Sowing of the following crops in June in light soils.

- Groundnut (bunch and spreading), Bajra
- Redgram, K-jowar, Setaria, H-jowar and other crop mixtures like *kharif* jowar in Redgram
- Groundnut + Redgram (3:1), Setaria + Redgram (2:1), Bajra +Redgram (2:1), similarly pulse crops in light retentive soils may be taken up.
- In *rabi* areas *i.e.*, medium black soils, sow Greengram/Blackgram, as a first crop in *kharif* to be followed by Safflower/*Rabi* Jowar/Bengal Gram/Wheat.

- When the land is kept fallow (Deep black soil) for *rabi* crops, compartment bunding or scooping or ridges and furrows or tied ridges for better soil and moisture conservation to be taken up. Harrowing after each rain and convert the plot into ridges and furrows or compartment bunds, which help in controlling weeds and conserving soil moisture may be taken up.

2. Normal onset of monsoon, but dry spells soon after germination:

- Protective irrigation for the crops sown wherever possible to be provided.
- Ratoon Bajra/Jowar for rejuvenation after rains.
- For crops like Groundnut, Urea spray immediately after rains for quick revival, wherever possible to be taken up.
- When the sown crops completely wither, Setaria, dolichos, Horse Gram, Matki, Cowpea and Sunflower soon after rains to be planted

3. No normal rains in June but onset of rains in July

- Groundnut (spreading), Sunflower and Setaria, in *kharif* area can be sown.
- Redgram, Sesamum, and Cowpea/Horsegram may be sown in light soils
- In *rabi* areas not to sow Greengram or Blackgram since it will delay *rabi* sowing.
- Harrowing to be done after each shower to remove weeds in *rabi* areas.
- Sowing of rainfed cotton by end of July to be taken up

4. Normal rains in August

- Complete sowing of Dryland cotton before the middle of August. Grow herbaceum cotton in place of hirsutum.
- Sunflower, Sesamum, Redgram, Setaria and Horsegram should be sown in light soils.
- Castor may be planted on contour bunds also.
- Relay cotton in groundnut in medium black soils can be taken up.

5. Normal rains in September

- Complete sowing of *rabi* jowar by middle of September in medium black soils of Northern taluks of Bijapur district. In the remaining taluks *viz*., Bagalkot, Hungund, Bilgi, Mudhol and Muddebihal complete *Rabi* Jowar sowing by first week of October. Early sowing of *Rabi* Jowar by 2 weeks than conventional sowing time in other district is preferred.
- Safflower as an entire crop to be sown before the end of September.
- Sunflower to be sown upto third week of September.

6. Sowing in October

- Continue sowing of *Rabi* Jowar till October 15th with 5% recommended level of fertilizer.
- Mixed cropping of *Rabi* Jowar+gram in 2:1 or 2:4 row proportion to be followed in Bijapur area.
- Safflower and bengalgram to be sown as mixed crops.
- The area under Safflower can be increased
- Bengalgram and Safflower can be sown in 2:4 row proportion for higher returns.
- Wheat may be sown till the end of October in retentive soils.
- Top dress *Rabi* Jowar with 10–15 kg N per hectare if adequate moisture is available in the soils

Mid-season Correction

Thinning: Thin out the population of *rabi* sorghum by removing every third row or alternate row within 40–45 days of sowing In later stages basal 3 to 4 leaves may be removed.

Soil cracks may be closed by repeated deep interculturing.

Mulching: Surface mulches of mixed trash or farm waste may be used wherever possible. Where farm waste is not available blade hoe may be used to form a thin layer of soil mulch to avoid cracks.

Alternate land use systems: Agri- horti sytem involving simarouba and guava based in medium black soil and tamarind based in shallow black soil & red soils enhances the productivity.

Table 69: Crop advisories and contingency models for *kharif* depending on rainfall or drought in Karnataka state

Main crop	15th June–15th July	16th–31st July	1st–15th August	16th–31st August	1st–15th September
Finger millet	*Use 15 to 20% higher seed rate with long duration varieties *viz.*, MR-1, MR-6, L-5. *Seed hardening (18 hrs soaking in water followed by 24 hrs shade drying) *Staggered nursery raising	*Re-sowing of the crop, if germination is failed *Dry sowing under insufficient moisture availability situations	*Intercultivation to remove weeds and create soil mulch *Make conservation furrow *Wherever large gaps are found plant seedlings after receipt of rains.	*Make dust mulch through intercultivation *Provide life saving irrigation, if possible	*Prepare fields for *rabi* for sowing of cowpea (KBC-2), Sunflower, Field bean, Horse gram (PHG-9) in September 1st fortnight and October 2nd fortnight

Main crop	15th June–15th July	16th–31st July	1st–15th August	16th–31st August	1st–15th September
		*Use 15 to 20% higher seed rate with medium and short duration varieties *Seed hardening *Thinning in already sown crop *If plant stand is too poor, remove seedlings and plant them on one side of the land after receipt of rains and go for sowing of little millet or Foxtail millet or horse gram as an alternate crop *Intercultivation to remove weeds and create soil mulch *Transplanting with the seedlings raised in staggered nursery	*Apply N-fertilizer immediately after receipt of rain as booster dose *Spraying with potassium		
Groundnut	*Sowing varieties: JL-24, S-206, R-8808, R-2001-3, ICP-8863, TS-3R, WRP-1, ICGV-91114, KCG-2	*Thinning and gap filling *Intercultivation to remove weeds and create soil mulch *If still unsown due to deficit rainfall, prefer spreading varieties (DSG-1, S-230)	*Intercultivation to remove weeds and create soil mulch *Earthing up and apply gypsum after receipt of rains *Life saving irrigation (micro irrigation) *Foliar spary of DAP (2%) Spraying of kaolin (4%)	*Intercultivation and use weed mulching *Provide life saving irrigation *Spraying of anti transpirant Kaolin (4%) *Removal of alternative rows for fodder purpose	*Life saving irrigation *Plan *rabi* crops Horse gram (October month) in light soils: Safflower, sorghum and jayadhar cotton (October) in black soils.
Maize	*Sowing composites (NAC-6002): Double cross hybrids (Ganga-11) hybrids (NAH-2049), NAH-1137)	*Gap filling *Re-sowing if plant stand is very poor *Make ridges and tied to conserve rain water	*Thinning and earthing up *Intercultivation to remove weeds *Apply 2, 4-D @ 500g ha^{-1} as post emergence at 30 days of growth	*Provide life saving irrigation *Apply kaolin Anti transpirant (4%)	*Harvest green cobs and remaining portion may be used for fodder

(Contd.)

Main crop	15th June–15th July	16th–31st July	1st–15th August	16th–31st August	1st–15th September
		*Top dressing only under sufficient soil moisture	*Provide life saving irrigation Top dressing after stress allevation (25 kg N ha^{-1}) *Spraying of anti transpirant kaoline (4%)	*Harvest lower leaves for fodder or as mulch *Top dressing after stress alleviation (25 kg N ha^{-1})	
Sorghum	*Sowing hybrids/varieties: CSH-5, CSH-9, CSV-4 and DSV-2	*Re-sowing within 15 DAS if population is less than 30% *Open conservation furrow for every 3–4 rows of crop *Frequent intercultivation to close soil crack and control weeds *If unsown, replace with pigeon pea/sunflower/groundnut (Spreading)	*Intercultivation and weeding *Thinning of plant population at 30–35 DAS or use as fodder/mulch	*Stripping of lower leaves and use as fodder or as mulch *Foliar spray of urea (2%)	*Harvest at physiological maturity and plan for early *rabi* crops Sunflower (KBSH-44, KBSH-53), Horse gram (PHG-9), Sesame (TMV-3, GT-1) and cowpea (IT-38956-1, TVX-944-02-E, KBC-2)
Pigeonpea/pearl millet/Sunflower	*Pigeon pea: TTB-7, BRG-2, ICP-7035, Pearl millet: ICTP-8203, ICMV-221, Asha, TS-3, Sunflower: KBSH-1, 41, 42, 44, 53	*Thinning of plant population in pearl millet and sunflower *Frequent intercultivation to close soil cracks and control weeds *In case of hand weeding, use weeds as mulch *Open conservation furrows at an interval of 1.5–2.0 m	*Remove every third row in pearl millet and use as fodder *Foliar spray of urea (2%) in sunflower *In *kharif* fallows, adopt ridges and furrows, tied ridges. *Compartmental, bunding to conserve rainwater for raising *rabi* crops later *Frequent intercultivation to close soil cracks and control weeds	*Harvest pearl millet for fodder purpose and allow ratooning *Repeated interculture and weeding *Foliar spray of KCl (2%) in sunflower *Spray anti-transpirant kaolin (4%)	*Plan for *rabi* crops like Horse gram (PHG-9), Foxtail millet, Sunflower (KBSH-1, 41, 42, 44, 53), Field bean (HA-3, HA-4), Cowpea (IT-38956-1, KBC-2) *Harvest for vegetable purpose in pigeonpea

Main crop	15th June–15th July	16th–31st July	1st–15th August	16th–31st August	1st–15th September
Green gram/ Black gram	*Green gram: Pusa Baisakhi, PDM-84, 178, S-4, China mung Black gram: Kharagaon-3, Rashmi, K-7, LBG-65	*Thinning and use as a mulch *In case of hand weeding, use weeds as mulch	*Foliar spay of urea (2%)	*Harvest at physiological maturity	*Harvest at physiological maturity or for fodder purpose *Plan for early *rabi* crops sorghum, safflower, chickpea, sesame, cowpea and horse gram
Fodder crops	*Sowing fodder bajra (Giant Bajra), Maize (South African Tall)	*Sowing fodder bajra (Giant Bajra), Maize (South African Tall) *Intercultivation to conserve moisture and avoid weeds	*Life saving irrigation if pond water is available *Harvest at flowering	*Plan for late *kharif* sowing of short duration finger millet (GPU-48), Horse gram (PHG–9), Cowpea (IT-38956-1, KBC-2) Field bean (HA-3, HA-4)	*Plan for late *kharif* sowing of short duration finger millet (GPU-48), Horse gram (PHG-9), Cowpea (IT-38956–1, KBC-2) Field bean (HA-3, HA-4) *Thinning and gap filling if already sown

Loss and damage assessment of drought

Droughts account for 85.8% of livestock losses and drought is the most lethal natural hazard to livestock (FAO, 2015). The distribution of drought-related losses show high relative losses in Sub-Saharan Africa but Central and South America, Southern Europe, the Middle East and Southern Australia are also at high risk (Natural Disaster Hotspots: A Global Risk Analysis, 2005). With the anticipated pressures on water resources and with more intense and severe droughts predicted, a paradigm shift is needed. A well-planned approach that focuses on reducing the impacts of drought is needed now.

Criteria for identification of drought affected areas

Various states have been following different norms for the declaration of drought.

Deficiency of rainfall

- Deficiency of rainfall of 25% and above in places where the annual rainfall is more than 1000 mm.
- 20% and above, where the annual rainfall is 750 mm–1000 mm.
- 15% and above where the annual rainfall is less than 750 mm.

Any two of the remaining norms

1. Compression/reduction in the cropped area of 50% and above under all principal crops.

2. Normal reduction in crop yields of 50% and above in comparison with average yields of preceding 5 years.
3. Dry spell and its impact on crops.

Calculation of Annual Drought

Data required

- Long term annual rainfall data
- First calculate mean of annual rainfall of the given station. It is also known as Normal annual rainfall of the given station

  Percentage Deviation of annual rainfall is calculated as

$$\%\text{ deviation of annual rainfall} = \frac{\text{Actual annual rainfall} - \text{Normal annual rainfall}}{\text{Normal annual rainfall}} \times 100$$

Note: Similarly as above %, deviations for seasons, months *etc.,* can be worked out

$$\textbf{Aridity Index} = \frac{\text{Water deficit}}{\text{Water need}}$$

$$\textbf{AI} = \frac{\text{Actual Evapo transpiration (AET)} - \text{Potential Evapo-transpiration (PET)}}{\text{Potential Evapo-transpiration (PET)}}$$

Aridity Anomaly Index (AI)

- This index is calculated on the basis of Thornthwait's water balance.
- This is the ratio of water deficit (PET-AET) to water need (PET).
- The departure of the index from the normal expressed as percentage of the normal is called Aridity Anomaly index.

Categories of drought based on aridity index

S. No.	Drought category	Anomaly value
1.	Mild drought	Up to 25%
2.	Moderate drought	26–50%
3.	Severe drought	More than 50%

Drought Signal

If MAI on any day is less than 0.75 it gives **ALARM**

If number of days with MAI < 0.75 is 10

↓

Gives signal 1 *i.e.,* **mild drought signal**

If number of days with MAI < 0.75 is 20

↓

If number of days with MAI < 0.5. is 10

↓

Gives signal 2 *i.e.*, **moderate drought**

If number of days with MAI <0.75 is 30

↓

If number of days with MAI < 0.50 is 20

↓

If number of days with MAI 0.25 is 10

↓

Gives signal 3 *i.e.*, **severe drought**

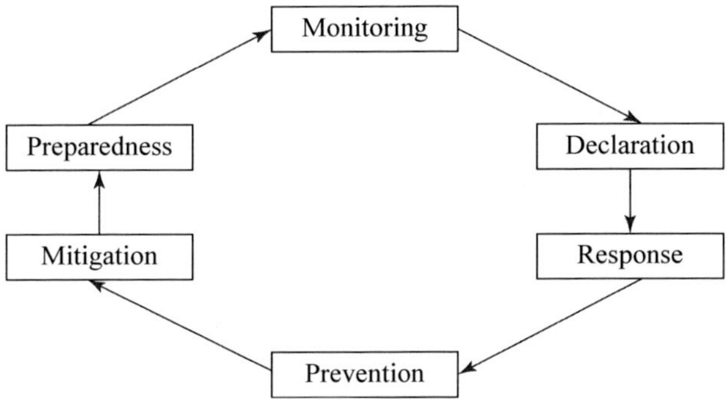

Fig. 17. A schematic description of a drought management cycle in India

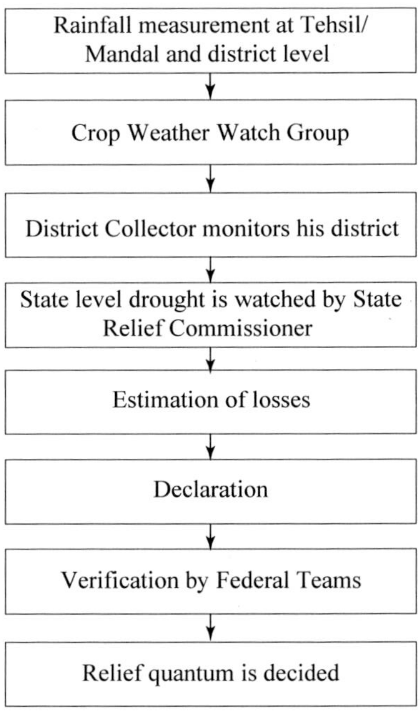

Fig. 18. A flow chart of drought declaration process in Indian states

Drought declaration in Karnataka

Karnataka state natural disaster monitoring center monitors rainfall, water level in reservoir and other relevant parameters on a daily basis in the rainy season and situation is reviewed on weekly basis by the development commissioner

- –20 to –59% moderate drought
- > 60% severe drought
- % area sown

On the basis of wide-ranging consultations with the meteorologists and agriculture scientists.

1. Rainfall deficiencies: If the total rainfall received during the months of June and July is less than 50% of the average rainfall.

If the total rainfall for the entire duration of the rainy season of the state, from June to September (the South-west monsoon) and or from December to March (North-east monsoon), is less than 75% of the average rainfall for the season.

2. The extent of area sown: Drought conditions could be said to exist if the sowing area of *kharif* crops is less than 50% of the total cultivable area by the end of July/August; for *rabi* crops-November/December.

3. Normalized difference vegetation index:

NDVI = (Near Infrared – Red/Near Infrared + Red)

It is necessary that the state declare drought only when the deviation of NDVI value from the normal is 0.4 or less.

However, the NDVI value needs to be applied in conjunction with other indicators and values.

4. Moisture adequacy index

MAI = (AE/PE)

where,

AE is Actual Evapotranspiration (%)

PE is Potential Evapotranspiration (%)

Table 70: Agricultural drought code (Based on moisture adequacy index) CAZRI

AE/PE (%) during different phenophases	Drought intensity	Phenophase growth wise code Seeding (S) Vegetative (V) Reprductive (R)		
76 to 100	No drought	S0	V0	R0
51 to 75	Mild drought	S1	V1	R1
26 to 50	Moderate drought	S2	V2	R2
25 or less	Severe drought	S3	V3	R3

Table 71: Intensity of agricultural drought

No drought	Mild	Moderate	Severe
S0V0R0	S0V0 R2	S0V0 R3	S0V2R3
S0V0 R1	S0V1R2	S0V1R3	S0V3R2
S0V1 R0	S0V2R0	S0V2R2	S0V3R3
S1V0 R0	S0V2 R1	S0V3R0	S1V2R3
S1V0 R1	S1V0 R2	S0V3 R1	S1V3 R3
S1V1R0	S1V1R1	S0V0R3	S0V2R3
S0 V1 R1	S1 V1 R2	S1 V1 R3	S2V0 R3

Source: Vittal and Rao CAZRI

These are recommended as the four standard monitoring tools which could be applied in combination for drought declaration.

Since the information on these indicators and indices are available at the level of taluka/tehsil/block, drought may be declared by the State Government at the level of these administrative units on the basis of observed deficiencies.

The system of estimating agricultural drought

Gujarat

- The Annawari procedure of estimating agricultural losses
- The committee estimates yield from three plots representing good, medium and poor crop conditions in a village, by actual harvesting.
- The Annawari score is calculated as, Annawari score = (12 × Observed yield/ Standard yield of the village) < 4 = Severe drought; 4 – 6 = Moderate drought

Maharashtra

- Paisewari system is followed,
- It works on a scale of 1 to 100 scores, and if a village gets less than 50 scores, it is declared drought-affected.

Andhra Pradesh

- Estimates of losses by remote sensing.

Chattisgarh

- Rainfall deficit itself is the criterion for declaring drought without actual estimation of agricultural losses.
- Normally, agricultural losses are estimated by random crop-cutting experiments at the district level and drought may be declared for the whole state or a part of it or even for a single district.
- Quantification of losses by the conventional revenue surveys has to wait for maturity of crops and delays the response. Modern methods of remote

sensing, GIS,GPS, modeling and real time communication can go a long way in hastening the responses.

- High-resolution spectral libraries and remote-sensing censors are evolving to improve precision and real timeliness of forewarning systems and estimation of losses.
- Karnataka has set up drought monitoring centers with the latest art of technologies and real-time communication. This experience is valid for other drought-prone states and for other countries of the region.

Drought mitigation strategies

- Policy support at national and state level
- Developmental funding to rural development and special assistance during natural calamities
- Input supply, access to credit and marketing and price support
- Farm advisory services

Non-Government Agencies like

- NGO, SHG, Philanthropic bodies and aid agencies

Major focus

- Education, building of awareness creation and community institutions and leadership
- Supplementing the Government effort in rural development

Short-term/immediate strategies

- Execution of labour-oriented schemes for employment generation and implementation of relief and development works-National Rural Employment Guarantee Act (NREGA) programme
- Good weather code-encashing production from good rainfall regions and managing low rainfall regions through transport of food grains from high production areas
- Establishment of Fodder/Seed/Grain banks
- Establishment of Cattle camps near water points

Long-term measures

- Long-range forecasting of rains (2–4 months in advance)
- Regional analysis of rainfall patterns
- Crop weather watch programs
- Value-added Agro-Advisories
- Integrated watershed development

- Land use diversification
- Water harvesting
- *In-situ* moisture conservation

Risk Management and Crop Insurance

Insurance Products from Agriculture Insurance Company of India Limited (AIC)

Agriculture Insurance Company of India Limited (AIC) was incorporated under the Indian Companies Act 1956 on 20th December, 2002 with an authorised share capital of INR 15 billion and paid up capital of INR 2 billion. AIC commenced business from 1st April, 2003.

- Pradhan Mantri Fasal Bima Yojana.
- The new Crop Insurance Scheme is in line with One Nation-One Scheme theme. It incorporates the best features of all previous schemes and at the same time, all previous shortcomings/weaknesses have been removed.
- Weather Based Crop Insurance Scheme (WBCIS).
- Government sponsored National Agricultural Insurance Scheme (NAIS) in operation since *Rabi* 1999–2000.
- Farm Income Insurance Scheme (FIIS) implemented on pilot basis in *Rabi* 2003–04 and *Kharif* 2004. Discontinued w.e.f. *rabi* 2004–05.
- Insurance linked to crop loan.
- Varsha Bima Yojana (Rainfall Insurance) being implemented by some insurance companies like ICICI-Lombard, IFFCO-Tokio, AIC on Pilot basis.
- Potato Crop Insurance.

Weather Index Insurance-Rainfall

Advantages

- Indemnity based on rainfall not on an individual
- Applicable to several crops
- Speedy settlement and transparency
- Weather indices could be deficit or excessive rainfall

Disadvantage

- Recurring droughts would make it more expensive

Contingent strategies for Livestock

Agriculture and animal husbandry are the integral part of the human existence. In fact, they always go together and do not live in isolation. Animal husbandry in India has started along with the civilization. Requirement of draught power for various agricultural operations and supply of nutritious milk and meat is mainly from livestock. In India, livestock population accounts for 512.06 million (as per Livestock census, 2012) with 165.4 million tonnes of milk and 7.38 million tonnes of meat (www.indiaagristat.com). By 2025, the demand for green and dry fodder is 1170 and 650 million tonnes, respectively against the estimated supply of 411.3 and 488 million tonnes. Thus, revealing the deficit of 64.87 and 24.92% of demand. While in Karnataka, the total livestock population is 27.7 million with 6.56 million tonnes of milk and 2.09 lakh tonnes of meat. The livestock population is dominant in the dryland agriculture which accounts for near 70% of the total cultivated area in the country. As a result of vicissitudes of monsoon, droughts and floods have become very common in different parts of the country affecting not only agriculture but also livestock production. Hence, considering contingency measures for the maintenance of livestock during droughts and floods becomes essential.

Suggested contingency measures for Drought
Before the event

Following measures to be taken to ameliorate the fodder deficiency

- Sowing of cereals (Sorghum/Bajra) and leguminous crops (Lucerne, Berseem, Horse gram, Cowpea) during North-East monsoon under dry land system for fodder production.
- Available Bajra/sorghum stover and groundnut haulms should be properly stored for future use.
- Encourage silage making with available maize fodder in the villages.
- Collection of groundnut haulms and groundnut cake for use as feed supplement during drought.
- Chopping of fodder should be made as mandatory in every village through supply and establishment of good quality chaff cutters.
- Harvesting and collection of perennial vegetation particularly grasses which grow during monsoon.
- Proper drying, bailing and densification of harvested grass from previous season. Creation of permanent fodder, feed and fodder seed banks in all drought prone villages.

During the event

- Harvest and use biomass of dried up crops (Bajra, Groundnut, Paddy, Maize, Green gram, Balckgram, Horse gram, Cowpea, soybean *etc.,*) material as fodder.

- Use of unconventional and locally available cheap feed ingredients especially groundnut cake and haulms as protein supplement for livestock during drought.
- Concentrate ingredients such as grains, brans, chunks & oilseed cakes, low grade grains *etc.*, unfit for human consumption should be procured from Govt. Godowns for feeding as supplement for high productive animals during drought.
- Continuous supplementation of mineral mixture to prevent infertility.
- Harvest the tree fodder (Neem, Subabul, Acasia, Pipal *etc.,*) and unconventional feeds resources available and use as fodder for livestock (LS).
- Available feed and fodder should be cut from Common Property Rights (Gomalas, Grasslands) and stall fed in order to reduce the energy requirements of the animals.
- Advise the farmers about the practice of mixing available kitchen waste with dry fodder while feeding.

After the event

- Training/educating farmers for feed & fodder storage.
- Maintenance/repair of silo pits and feed/fodder stores.
- Encourage progressive farmers to grow fodder crops of sorghum/bajra/maize (UP chari, MP chari, HC-136, HD-2, GAINT BAJRA, L-74, K-677, Ananad/ African Tall *etc.*,) on their own lands & supporting them with assisting infrastructures like seeds, manure.
- Supply of quality fodder seed (multi cut sorghum/bajra/maize varieties) and fodder slips of napier, guinea grass well before monsoon.
- Replenish the feed and fodder banks.

Suggested contingency measures for Cyclone

Before the event

- Harvest all the possible wetted grain (bajra, groundnut, paddy, maize, green gram, black gram, horse gram, cowpea, soybean *etc.,*) and use as animal feed.
- Stock of anti-diarrheal drugs and electrolytes should be made available for emergency transport.
- Don't allow the animals for grazing in case of early forewarning (EFW) of cyclone.
- In case of EFW of severe cyclone, shift the animals to safer places.

During the event

- Treatment of the sick, injured and affected animals through arrangement of mobile emergency veterinary hospitals/rescue animal health workers.

- Diarrhea out break may happen. Health camps should be organized.
- In severe cases un-tether or let loose the animals.
- Arrange transportation of highly productive animals to safer place.
- Spraying of fly repellents in animal sheds.

After the event

- Repair of animal shed.
- Deworm the animals through mass camps.
- Vaccinate against possible disease out breaks like HS, BQ, FMD and PPR.
- Proper dispose of the dead animals/carcasses by burning/deep burying (4–8 feet) with lime powder (1kg for small ruminants and 5 kg for large ruminants) in pit.
- Bleach/chlorinate (0.1%) drinking water or water resources.
- Collect drowned crop material, dry it and store for future use.
- Sowing of short duration fodder crops in unsown and water logged areas when crops are damaged and no chance to replant.
- Application of urea (20–25 kg/ha) in the inundated areas and CPR's to enhance the bio mass production.

Suggested contingency measures for Health and Disease management

Before the event

- List out the endemic diseases (species wise) in that district.
- Procure and stock emergency medicines and vaccines for important endemic diseases of the area.
- All the stock must be immunized for endemic diseases of the area.
- Surveillance and disease monitoring network to be established at Joint Director (Animal Husbandry) office in the district.

During the event

- Constitution of Rapid Action Veterinary Force.
- Performing ring vaccination (8 km radius) in case of any outbreak.
- Restricting movement of livestock in case of any epidemic.
- Rescue of sick and injured animals and their treatment.

After the event

- Conducting mass animal health camps.
- Conducting fertility camps.
- Mass deworming camps.

Suggested contingency measures for drinking water

Before the event

- Identification of water resources.
- Rain water harvesting and create water bodies/watering points (when water is scarce use only as drinking water for animals).

During the event

- Restrict wallowing of animals in water bodies/resources.

After the event

- Bleach (0.1%) drinking water/water *Source*s. Provide clean drinking water.

Suggested contingency measures for Insurance

Before the event

- Encouraging insurance of livestock.

During the event

- Listing out the details of the dead animals.

After the event

- Submission for insurance claim and availing insurance benefit.
- Purchase of new productive animals.

Contingent strategies for Poultry

Poultry rearing is an integral component of the farming system approach providing meat and egg to the both vegetarian and non-vegetarian population. In the country, there are 729.21 million poultry birds with the production of 88138.57 million eggs (www.indiaagristat.com). In Karnataka, there are 53.44 million poultry birds with the production of 5067.14 million eggs. Successful poultry production needs intensive care and maintainance in terms of nutrition, drinking water and health on a regular basis. During exigencies of droughts and floods, further care and management is required on a war footing to save the life and continue the production.

Suggested contingency measures for Drought

Shortage of feed ingredients

Before the event

- Storing of house hold grain like maize, broken rice *etc.*, into use as feed in case of severe drought.

During the event

- Supplementation only for productive birds with house hold grain.
- Supplementation of shell grit (calcium) for laying birds.
- Culling of weak birds.

After the event

- Supplementation to all survived birds.

Shortage of drinking water

During the event

- Use water sanitizers or offer cool hygienic drinking water.

Health and disease management

Before the event

- Culling of sick birds.
- Deworming and vaccination against RD and IBD.

During the event

- Mixing of Vit. A,D,E, K and B-complex including vit C in drinking water (5ml in one litre water).

After the event

- Hygienic and sanitation of poultry house.
- Disposal of dead birds by burning/burying with lime powder in pit.

Suggested contingency measures for Cyclone

Shortage of feed ingredients

Before the event

- In case of EFW, shift the birds to safer place.
- Storing of house hold grain like maize, broken rice, bajra *etc.*
- Culling of weak birds.

During the event

- Use stored feed as supplement.
- Don't allow for scavenging.
- Protect from thunder storm.

After the event

- Routine practices are followed.

Shortage of drinking water

- Use water sanitizers or offer cool hygienic drinking water.

Health and disease management

Before the event

- In case of EFW, add antibiotic powder in drinking water to prevent any disease outbreak.

During the event

- Sanitation of poultry house.
- Treatment of affected birds.
- Prevent water logging surrounding the sheds.
- Assure supply of electricity.
- Sprinkle lime powder (5–10 g per square feet) to prevent ammonia accumulation due to dampness.

After the event

- Disposal of dead birds by burning/deep burying with lime powder in pit.
- Disposal of poultry manure to prevent protozoal problem.
- Supplementation of coccidiostats in feed.
- Vaccination against Ranikhet Disease (0.5 ml S/c).

Contingent strategies for Fisheries

Inland fisheries is one of the components of integrated farming systems in both dryland and coastal ecosystems. In the country, total inland fish production accounts for 9.71 million tonnes, while in the state of Karnataka its production is 1.97 lakh tonnes (www.indiaagristat.com). Quite oftenly its successful cultivation is very much threatened due to aberrant weather conditions including droughts, cyclones and floods. Special care and management before the event will reduce the damage substantially.

Suggested contingency measures for Drought
Capture Inland Fisheries

Shallow water depth due to insufficient rain/inflow

Before the event

- Observe water level. Advice fishermen to harvest as much as possible fish live stock.

During the event

- Harvest the complete fish live stock.

After the event

- Report the loss to Revenue & Fisheries Dept.

Changes in water quality
Before the event

- Observe water quality like dissolved Oxygen & pH.

During and after the event

- Report the matter to Revenue & Fisheries Dept.

Any other

- To explore the possibility of shifting the livestock to other water resources.

B. Aquaculture

Shallow water in ponds due to insufficient rain/inflow

Before the event

- Observe water level. Advice for fishermen to harvest maximum fish live.

During the event

- Addition of water, lime for tackling salt load.

After the event

- Report the loss to Revenue & Fisheries Dept.

Impact of salt load build up in ponds/change in water quality

Before the event

- Observe water quality like dissolved Oxygen & pH.
- Addition of water, lime for tackling salt load.

During and after the event:

- Report the matter to Revenue & Fisheries Dept.

Any other

- To explore the possibility of shifting the live stock to other water resources.

Chapter 8

Rainwater Harvesting

Successful crop production both in the rainfed and irrigated ecosystems depends on timely onset of monsoon, its distribution during the crop life cycle and collection of water in various storage structures *viz;* tanks, reservoirs and ground water table. It is therefore important that adequate supplies of water should be developed to sustain life. Development of water supplies should, however, be undertaken in such a way as to preserve the hydrological balance and the biological function of our ecosystems.

Consequently, the human endeavour in the development of water sources must be within the capacity of nature to replenish and to sustain. If this is not done, costly mistakes can occur with serious consequences. The application of innovative technologies and the improvement of indigenous ones should therefore include management of the water resources to ensure sustainability and to safeguard the source against pollution.

As land pressure rise, cities are growing vertical and in country side more forest area encroached and being used for agriculture. In India small farmers depends on monsoon where rainfall is from June to October and much of the precious water is soon lost as surface runoff. While irrigation may be the most obvious response to drought, it has proved costly and can only benefit a fortunate few. There is now increasing interest in the low cost alternative-generally referred to as "Rain Water Harvesting (RWH)".

Water harvesting is the activity of direct collection of rainwater, which can be stored for direct use or can be recharged into the ground water. Water harvesting is the collection of runoff for productive purposes.

Rivers, lakes and ground water are secondary source of water. In present times, we depend entirely on such secondary source of water. In the process, it is forgotten that rain is the ultimate source that feeds all these secondary sources and remain ignorant of its value. Water harvesting is to understand the value of rain, and to make optimum use of rainwater at the place where it falls.

What is Rainwater?

Rainwater is water that falls on a roof or other hard, impervious non-pedestrian, non vehicular surface and is captured, filtered and stored. Further, it is the water that falls in the cultivated land, stored in the effective root zone and used consumptively for crop production. Upon saturation it enters the ground water.

What is a Rainwater Harvesting System?

Rainwater harvesting (RWH) is the accumulation and storage of rainwater for later use. Possible uses include agricultural production in arid and semiarid regions, domestic supply, and environment measures, *e.g.,* the refilling of aquifers. Rainwater harvesting systems have been developed primarily in area where rainfall is not sufficient to support crop production or where water sources were scarce, of poor quality or inaccessible *e.g.,* due to very deep ground water. Numerous historic examples of RWH exist throughout the world.

The Water Cycle

Every living creature and vegetation must have water in order to survive. Only 3% of all water on earth is fresh water and 68% of water is frozen ice on the north and south poles. All humans and most animals as well as nearly all vegetation can only survive on fresh water free of salt and minerals. Water for irrigation must also be fresh water and applied sparingly by flood, furrow or drip irrigation, otherwise the irrigated soil will turn saline and unproductive for many years to come.

The others 97% of the water on planet earth is saline sea water in which whales, fish, corals and plankton flourish. Slightly salty water may be used for watering livestock and other animal, although fresh water is healthier.

Nearly half of all deep hole borewell are dry or contain saline (brackish) water with mineral harmful to humans, animals and plants. Where fresh water is pumped up from very deep bore holes, it is called mining fossil water because the water withdrawn cannot be replaced by rains. Therefore, there are only two sustainable solutions to the world's increasing demand for water, namely rainwater harvesting and water conservation.

Freshwater sources are replenished in a water cycle through the following activities

1. Evaporation is an almost invisible vapour rising from water surfaces.
2. Transpiration is evaporation from vegetation and soil surfaces.
3. Precipitation from clouds falls as fogs, mist, rains, hails and snow.
4. Rain water runoff is rainwater running off all kind of surfaces.
5. Drainage is ability to drain excess water away from catchments.
6. Infiltration is the movements of water into the soil from the surface.
7. Percolation is the movement of water through the soil to the underground layers.

8. Permeability is the rate at which water penetrates through soils down into the underground water table.
9. *In-situ* storage *is* storage of water in the voids between particles of soil and sand.
10. Subsurface flow is a flow of water in the voids of soil and sand particles.

Resource Conservation in field

The first steps involved in resource conservation are:

- Use practices based on the existing traditional systems.
- Adopt contour bunding in new areas being brought under plough and with large farm plots.
- Strengthen the existing bunds across the slopes and provide weirs.
- In the cases of drainage line treatment, start from ridge line to the bottom.
- The traditional loose boulders structure as well as grassing may be adopted.
- Later, gully plugging and construction of small check dams on the drainage line are carried out.
- After bunding and provision of water way, grow dual purpose plant species on the bunds.
- Encourage farmers to adopt contour farming.
- Provide small cross section bunds with a small furrow on the upper side, preferably with a heavy country plough or mould board plough.
- The water course is conveyed with vegetation, preferably before other treatment are superimposed.

Further steps are

- Consider the wells existing in the area and then calculate the potential water supply as well as the increased supply through the various field and drainage line treatments.
- Follow strict water budgeting for sustainable use of the harvested rainwater.
- Allow only low duty crops. Avoid sugarcane, rice and wheat. Encourage pulses and oil seeds.
- Introduction of horticulture in class V and above lands. Provide irrigation cover for first two years. Do not irrigate the fruit trees without providing suitable basins.

Rainwater budgeting

The total amount of rainwater falling in the watershed area and the amount of outflow from the watershed have to be accounted properly to help in deciding the number and the capacity of each conservation and storage structure. This type of accounting of rainwater received in the watershed, possible storage in different structures and the outflow from the watershed is called "rain water budgeting".

In low rainfall areas (upto 750 mm) most of the runoff can be collected and stored in farm ponds, nala bunds, check dams, trenches and percolation tanks. Nevertheless, keeping in view of the sustainability of the storage structure like tanks, sizable amount of rainwater has to be allowed as runoff. From the experience of Dry land research project at GKVK, Bangalore and Dryland Research Station at Bijapur, it is recommended to provide a farm pond of 250 m^3 capacity for every hectare of cultivated catchment in red soils areas and 150 m^3 capacity in low rainfall black soil areas. If the land is uncultivated and is a forest catchment, trenches of suitable size for harvesting rainwater may be created. In general, about 25% of the total runoff may be retained in the watershed allowing 75% of runoff into natural streams and tanks in the lower areas.

In high rainfall areas as in zone 9 and 10 (Hilly and Coastal zone) it is not feasible to collect 25% of the runoff due to huge costs of creating storage structure. If the storage is for protective irrigation, the amount required to provide 3–4 irrigation to the specified area may be stored. In general, strategy in high rainfall areas should be to open trenches across the slope and to create percolation tanks to increase the ground water recharge.

As a thumb rule, about 5 per cent of the total rainfall enters into water table and recharge ground water. It will however be higher when recharge structures are created. In any case, withdrawal from open wells as well as bore wells should not exceed the annual recharge. Otherwise an imbalance will be created.

Example: The total amount of water available as runoff and groundwater recharge is calculated in Ankanahalli Watershed

I. Total amount of runoff

1. Area of the watershed 866 ha
2. Annual rainfall 750 mm
3. Number of run-off producing rains (of more than 20 mm/day) 11 days
4. Average total annual precipitation of the above run-off producing rains) (10 years average) 418 mm
5. Assuming 25% as average run-off of the runoff causing rains, the runoff will be 104.5 mm
6. Volume of run off from the watershed = Area × depth of run-off

$$= \frac{866 \times 104.5 \times 10000}{1000}$$

$$= 904970 \text{ cum}$$

= 90.45 ha.m or 905.0 thousand cubic meter (TCM)/year

= 905 TCM/year

Note: Structures to retain about $^1/_4$ of the runoff or 225 TCM of runoff at a time may be created. It comes to around 3.5% of the rainfall in the present example

II. Groundwater recharge

1. Total area of the watershed = 866 ha
2. Average rainfall of the area = 750 mm
3. Total recharge @5% of the total rainfall $= \dfrac{\text{Area} \times RF \times 5}{100}$

$$= \frac{866 \times 750 \times 5 \times 10000}{1000 \times 100}$$

$$= 324750 \text{ cu m}$$

$$= 325 \text{ TCM/year}$$

Withdraw from the groundwater in the watershed should not exceed 325 TCM

Science of Water Harvesting

In scientific terms, water harvesting refer to collection and storage of rainwater and also other activities aimed at harvesting surface and ground water, prevention of losses through evaporation and seepage and all other hydrological studies and engineering inventions, aimed at conservation and efficient utilization of the limited water endowment of physiographic unit such as a watershed.

Rain is a primary source of water for all of us. Technologies can be divided in two main areas:

1. *In-situ* rain water harvesting
2. *Ex-situ* rainwater harvesting

Who can harvest rainwater ? and where?

- People planning construction of house, modification of house, existing house, *etc.,* from rooftops.
- Govt. Buildings, Institutions, Hospitals, Hotels, Shopping malls *etc.*, from rooftops and open areas.
- Farmlands, public parks, playground, *etc.*
- Paved and unpaved areas of a layout/city/town/village.

Need for Rainwater Harvesting

1. As water is becoming scarce, it is the need of the day to attain self–sufficiency to fulfill the water needs.
2. As urban water supply system is under tremendous pressure for supplying water to ever increasing population.
3. Groundwater is getting depleted and polluted.
4. Soil erosion resulting from the unchecked runoff.
5. Health hazards due to consumption of polluted water.

Advantages of Rainwater harvesting

1. The water is free-only cost for collection and use.
2. Environment friendly and easy approach for water requirements.
3. RWH is the ideal solution for all water requirements.
4. The rooftops rainwater harvesting is less expensive, easy to construct, operate and maintain.
5. Increasing in ground water level.
6. Mitigates the effects of drought.
7. Reduces the runoff, which other wise flood storm water drains.
8. Reduces flooding of roads and low-lying area.
9. Reduced soil erosion.
10. Improve the ground water quality.
11. Superior quality water for landscape irrigation.
12. In desert, where rainfall is low, rainwater harvesting has been providing relief to people.
13. Low cost and easy to maintain.
14. Reduces water and electricity.
15. To meet water demand for domestic, animal and recreational use.
16. To provide life saving/supplemental irrigation to crops and plantation.
17. To spray insecticides/pesticides in crop management.

in-situ Rainwater Harvesting

Methods that increase the amount of water stored in the soil profile by trapping or holding the rain water where it falls is called *in-situ* water harvesting. It is also called as soil and water conservation.

in-situ rainwater harvesting is basically a prevention of net runoff from a given cropped area by holding rain water and prolonging the time for infiltration. This system works better where the soil water holding capacity is large enough and the rainfall is equal or more than the crop water requirement, but moisture amount in the soil is restricted by the amount of infiltration and or deep percolation.

Typically, the *in-situ* systems do not cover a distance of more than 5–10 m from the point of water intake to the point of infiltration into the storage medium, mostly the soil. Often, *in-situ* RWH primarily serves the recharge of soil water for producing crops and other vegetation and reducing the need for irrigation.

in-situ harvesting of rain water in farmland increases the yield of crops and will often determine whether there will be anything to harvest at all. Most farmers know and apply some of the techniques of soil conservation that make rainwater percolate into the soil instead of eroding the farmland. Among several techniques the ones, which are recommended and followed predominantly are discussed in the chapter on soil and moisture conservation.

I. Recharging ground water aquifers

Excavation of the bore wells for drinking as well as for irrigation has become very common in the dryland and irrigated agriculture. Digging of bore wells to a depth of 1500–2000 feet below the surface is observed in the dry areas of Karnataka *viz.* Kolar, Chikkaballapur, Bengaluru Rural, Ramanagara, Tumakuru and Chitradurga. These bore wells may yield water or found dry totally. Lifting of water from such a deep level involves enormous energy and many a times quality of water both for drinking and agriculture is found to be not suitable. Although, there is a need for registering in the ground water board before excavation and minimum distance of 250 m between two bore wells to be maintained but, excavation is going on indiscriminately. As a result, many taluks have been declared as grey or black areas due to over exploitation more than the ground water recharge. Hence, recharging of ground water should be made compulsory. Ground water aquifers can be recharged by various kind of structures to ensure percolation of rainwater in the ground instead of draining away from the surface. Commonly used recharging methods are:

(a) Recharging of borewells
(b) Recharging of dugwells
(c) Recharge pits
(d) Recharge trenches
(e) Soak ways or Recharge Shafts
(f) Percolation tanks

A. Recharging of borewells

Rainwater collected from rooftop of the building is diverted through drain pipes to settlement or filtration tank. After settlement filtered water is diverted to borewells to recharge deep aquifers. Abandoned borewells can also be used for recharge.

Optimum capacity of settlement tank or filtration tank can be designed on the basis of area of catchment, intensity of rainfall and recharge rate. While recharging, entry of floating matter and silt should be restricted because it may clog the recharge structure. "first one or two shower should be flushed out through rain separator to avoid contamination. This is very important, and all care should be taken to ensure that this has been done".

Recharging of borewell with runoff water

In the agricultural fields, excess runoff water from the catchment area or field drain is diverted to the recharge pit of borewell. A pit around the casing pipe with 3m × 3m × 2.85m dimensions is excavated and different layers are filled as indicated in the Fig. 19.

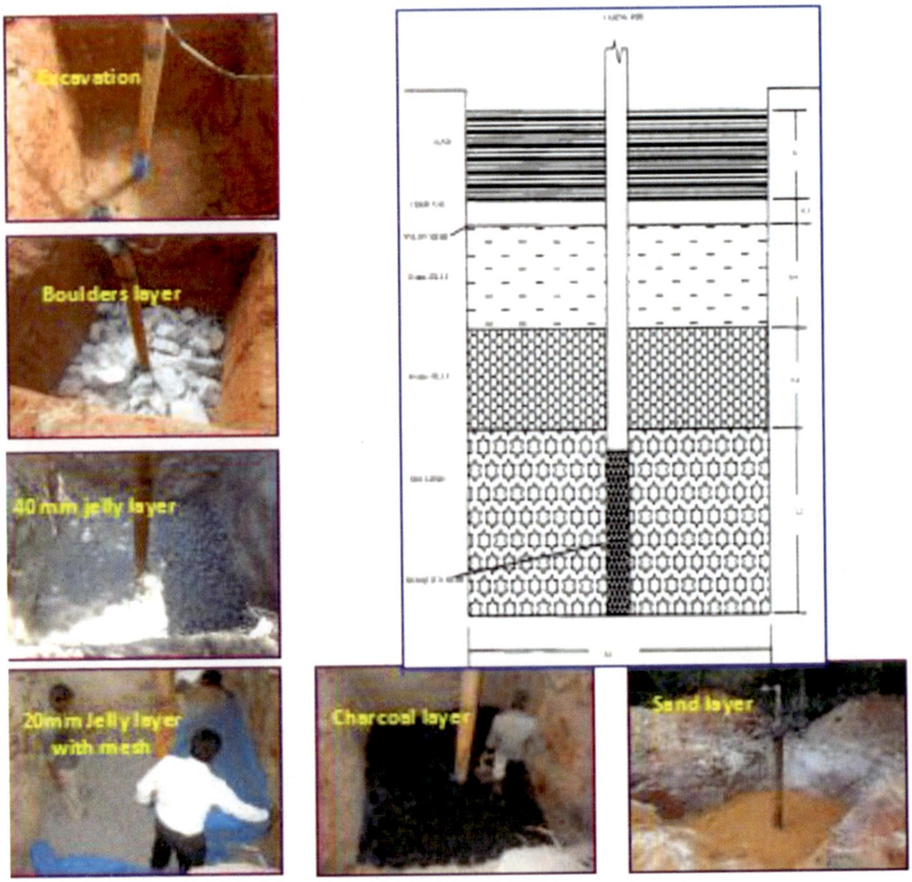

Fig. 19. Schematic diagram of borewell recharge with different layers

Fig. 20. Ground water recharge model

Particulars	Borewell
Location	Hosapalya, Nelamangala
Depth	250 ft
HP of pump	5
Type of pump	Multi stage submersible
Filter dimensions	Length: 3.0 m Width : 3.0 m Depth : 2.9 m
Catchment	1 ha

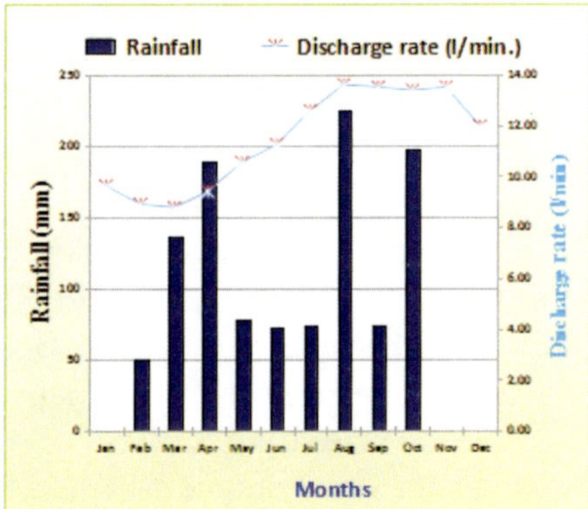

Fig. 21. Recharging of borewell with runoff water in Hosapalya village

After implementing recharge treatment, the average discharge rate of bore well with filter bed was 11.4 l/min throughout the year and in the rainy season discharge rate was 12.8 l/min and in summer season it was 9.4 l/min. The discharge rate was increased during the commencement of rainy season and declined towards the cessession of rainfall.

B. Recharging of dugwells

Dug well can be used as recharge structure. Rainwater from the rooftop is diverted to dug wells after passing it through filtration bed. Cleaning and desilting of dug well should be done regularly to enhance the recharge rate. The filtration method suggested for bore well recharging could be used.

C. Recharge pits

Recharge pits are small pits of any shape rectangular, square or circular, constructed with brick or stone masonry wall with deep hole at regular interval. Top of pit can be covered with perforated covers. Bottom of pit should be filled with filter media.

The capacity of the pit can be designed on the basis of catchment area, rainfall intensity and recharge rate of soil. Usually the dimensions of the pit may be 1–2 m width and 2–3 m deep depending on the depth of pervious strata. These pits are suitable for recharging of shallow acquifers, and for small houses.

D. Recharge trenches

Recharge trench is provided where upper impervious layer of soil is shallow. It is a trench excavated on the ground and refilled with porous media like pebbles, boulder or brickbats. It is usually made for harvesting the surface runoff. Borewells can also be provided inside the trench as recharge shafts to enhance percolation. The length of the trench is decided as per the amount of runoff expected. This method is suitable for small houses, polyground, parks and roadside drains. The recharge trench can be of size 0.50–1.0 m wide and 1.0–1.5 m deep.

E. Soak way or Recharge Shafts

Soak way or recharge shafts are provided where upper layer of soil is alluvial or less pervious. These are bored hole of 30 cm diameter up to 10–15 m deep, depending on depth of pervious layer. Bore should be lined with slotted or perforated PVC/MC pipe to prevent collapse of the vertical sides. At the top of soak away required size sump is constructed to retain runoff before the filters through soak away. Sump should be filled with filter media.

F. Percolation tanks

Percolation tanks are artificially created surface water bodies, submerging a land area with adequate permeability to facilitate sufficient percolation to recharge the groundwater. These can be built in big campuses where land is available and topography is suitable. Surface runoff and roof top water can be diverted to this tank. Water accumulating in the tank percolates in the soil to augment the ground water. The stored water can be used directly for gardening and raw use. Percolation tanks should be built in gardens, open spaces and roadside green belts of urban area.

II. Storage of harvested rain water

ex-situ technologies are defined as systems which have separated catchment areas leading the water to the point of water storage. The most common *ex-situ* system is the rooftop rainwater collection.

ex-situ systems include catchment areas with usually low or little infiltration capacities like, *e.g.*, rooftops, roads or pavements; yet also bits of ground or rocks are used. These catchment areas allow for a fairly easy collection of substantial amounts of water. This is then stored in wells, dams, ponds or cisterns and, when needed, abstracted and distributed for irrigation or domestic, public and commercial uses.

ex-situ rainwater harvesting can be made in the excavated farm pond and watersheds may retain 25–30% of runoff and allow the remaining to flow through the streams. On a macro level estimation, taking 5–7% of total rainfall about 6.32 m ha-m of water can be harvested annually in the low rainfall areas in India (Table 72).

Table 72: Estimated volume of rain water for small scale water harvesting structures in India

Rainfall zone (mm)	Geographical area (m ha)	% rainfall for effective storage structures	Harvestable runoff in storage structures (m ha-m)
100–500	52.07	5	0.78
500–750	40.26	6	1.51
750–1000	65.86	7	4.03
		Total	**6.32**

Harvested water can be efficiently used for multiple purposes including protective irrigation to crops in the command area.

ex-situ rainwater harvesting consists of 5 components

1. Rainfall.
2. Catchment areas, also called watersheds, onto which the rainwater falls.
3. Gutters, or conveying channels, to bring rainwater from a catchment area to storage reservoir.
4. Storage reservoirs can be tanks, ponds, dams *etc.*
5. Retrieval water is extracted from reservoirs either by gravity or by pumps and lifts.

Types of *ex-situ* Rainwater Harvesting

1. Farm pond
2. Check dams
3. Nala bunds
4. Roof water harvesting

1. Farm pond

Farm Pond is a dug out structure with definite shape and size having proper inlet and outlet structures for collecting the surface runoff flowing from the farm area. It is one of the most important rain water harvesting structures constructed at the lowest portion of the farm area. The stored water must be used for irrigation only.

Farm ponds have a significant role in rainfed regions where annual rainfall is more than or equal to 500 mm.

- If average annual rainfall (AAR) varies between 500–750 mm, the farm ponds with capacity of 250–500 m^3 can be constructed.
- If AAR is more than 750 mm, the farm ponds with capacity more than 500 m^3 can be planned particularly in black soil regions without lining.

Types of farm pond

As per the method of construction and their suitability to different topographic conditions, there are three types of farm ponds, they are:

1. Excavated farm pond for flat topography
2. Excavated cum embankment ponds in mild sloping topography
3. Embankment farm ponds for hilly and rugged terrain.

Selection of site

Selecting the suitable site is considered as one of the most important steps in planning for farm ponds. It depends on,

- Local soil condition
- Topography of area
- Drainage capacity
- Infiltration
- Rainfall pattern and distribution

Location: The selection of a site for a farm pond is critical in maximizing its storage capacity. The pond must be located in a corner of a plot of land so that it does not disturb farm operations like ploughing. It must be located at least 3 m away from other farmers' field common lands. The slope of the land and the slope's direction must also be carefully evaluated.

Excavation: A test pit is dug out before finalizing the location and depth of excavation. The excavation and transportation of earth can be accomplished with a combination of manual labour or with machines like excavators and tractors.

For 1 ha of catchment area, 250 m^3 volume of farm pond is recommended.

Fig. 22. Farm pond brick lining

Construction Method: Material required: Soil, Cement, LDPE (400 micron) sheet, Bricks.

Excavate the earth by 12 m × 12 m top length, 6 m × 6 m bottom length and the vertical height is 3 meter, in 1:1 slope. Lay the 400 micron LDPE sheet all around the farm pond and bottom such way that the sheet should not be damaged by minor stones and boulders or any sharp objects and lay the bricks all around the farm pond with cement to control seepage and percolation losses. In 250 m^3 capacity farm pond 2,50,000 litres of water can be stored at any given of time. Depending upon the intensity of rainfall and runoff causing rains, there can be filling of water in the farm pond 2–3 times in the rainy season. Hence through sprinkler system the entire catchment area of one hectare can be irrigated with 2.5 cm depth of water every time pond is filled atleast during critical stages.

In case of farm ponds, area at the top, at half the depth and at the bottom are obtained by measuring the sides and substituted in the formula below for getting volume (m^3) of water stored.

$$V = \frac{A + 4B + C}{6} \times D$$

V is volume of pond in cubic meters (m^3)

A is top area of the pond (m^2)

B is area at half the depth of pond (m^2)

C is the bottom area of pond (m^2)

D is depth in meters (m)

If the ponds are trapezoidal in shape V is arrived by using the formula

$$V = \frac{D}{3}\left[A + B + \sqrt{A + B}\right]$$

Where V is volume, d is depth, A is top area and B is bottom area

Table 73: Seepage and Percolation losses in soil

S. No.	Types of Soil	Water loss through seepage (Cumecs/million m^2 of wetted area)	Drop in depth per day (cm)
1.	Heavy clay loam	1.21	10.36
2.	Medium clay loam	1.96	16.84
3.	Salty clay Loam	2.86	24.61
4.	Sandy loam	5.12	44.03
5.	Loose sandy soil	6.03	51.80
6.	Porous gravelly soil	10.54	90.65

Water is precious anywhere on earth but more so in low rainfall semi-arid regions. While we make all efforts to store runoff water, equal efforts are needed to identify cost effective sealants and efficient use of stored water for purposes like nursery raising, growing high value horticultural crops, fish rearing and adoption of drip irrigation for fruit crops.

Advantages of Farm Ponds

- Stored water can be used under extreme situations for drinking by animals and spraying of plant protection chemicals.
- They provide water to start growing crops, without waiting for rain.
- They provide protective irrigation water during dry spells between rainfalls. This increases the yield, the number of crops in one year, and the diversity of crops that can be grown.
- Bunds can be used to raise vegetables and fruit trees, thus supplying the farm household with an additional *Source* of income and nutritious food.
- Farmers are able to apply adequate farm inputs and perform farming operations at the appropriate time, thus increasing their productivity and their confidence in farming.
- They check soil erosion and minimizes siltation of waterways and reservoirs.
- They supply water for domestic purposes and livestock.

- They promote fish rearing.
- Cultivation of Azolla for cattle feeding which increases the milk yield.

Nourishing horticultural and vegetable crops using farm pond during dry spells

S. No.	Name of the farmer	Vegetables	Yield (kg)	Income (₹)
1.	Mr. Ravikumar Mudalapalya	Drumstick	33.0	1000.00
		Curry leaf	30.00	300.00
		Pudina	25.00	500.00
2	Mr. Krishnappa Chikkaputtaiahna Palya	Tomato	50.0	3500.00
		Leafy vegetables	40.0	2000
3.	Mr. Gubbanna, Chikkamaranahalli	Nourishing Mango, Drumstick and Amla which are grown around the farm pond		

Farm pond

Fish rearing

Cultivation of Azolla

Protective irrigation through drips and sprinkler

Piped irrigation

Traditional water bodies

Based on the local wisdom, communities have designed effective and efficient methods for storing the rain water in different states. Rajasthan having the highest area under dryland agriculture, every effort is being made to harvest rainwater both in urban and rural areas. Some of the novel systems prevalent are Baori and Jhalara (step wells), Nadi (village pond), Tanka (cistern), Khadin (run off farming system) are adopted. Baori and Jhalara are largely to benefit the urban and semi urban population whereas Nadi, Tanka and Khadin are well suited for the rural population. The dependancy of drinking water in villages in Western Rajasthan is 42.4% on Nadi, 34.7% on Tanka, 15.0% on wells and tube wells and 7.8% on other sources. Thus suggesting that the rain water harvesting is the backbone of drinking water supply in rural areas. The storage of rain water in underground cistern, locally known as tanka is a common practice in this region. Various types of tanka from rectangular to circular, in capacity from as small as 1000 litres–5,00,000 litres are prevalent in this region. The construction of these tanka also varies from simple mud plaster to lime mortar, cement concrete, ferro cement, fibre glass and PVC. Central Arid Zone Research Institute, Jodhpur has perfected the technology of tanka construction for various types of users.

(a) Nadi: is a dugout pond used for storing runoff water available from adjoining natural catchment during the rainy season. Generally, Nadi have the limitation of high evaporation losses due to large exposed surface area, high seepage losses through porous sides and bottom which can be reduced with LDPE lining.

(b) Khadin: is a unique practice of water harvesting and moisture conservation, is suitable for deep black soils surrounded by some sort of natural catchment.

(c) Talapariges: The small traditional water bodies of Karnataka that saved many lives during drought years barely quarter century ago are forgotten by and large forgotten today. Though talaparige looks like a small tank, it is neither a tank nor a well. For a casual onlooker hurried look, it looks like a small pond. Not deep or wide as an open well. It is a unique structure. It is the only source of irrigation for summer crop. Water is taken through a channel and by gravity-without using a motor. Talapariges are situation specific and found in the areas of shallow topsoil beneath which there is a rocky stratum. The rainwater that falls on the land and the nearby hills slowly gets percolated in the topsoil. The rocky stratum doesn't allow water to go very deep. As such, it finds out a lateral route. This low–pressure flow of water comes up in some points where soil is pretty sandy. Our wise elders used to identify such marshy spots and develop it into a water body. To avoid the surrounding soil collapsing into the pond, stonewall was built. The pond with round-shaped retaining wall is called *gerandi bavi* and the ones with square walls as *chowka bavi*. Protection wall serves another purpose too. It prevents the 'eye of water' getting buried when the tank is filled with water. The retaining wall served two purposes; one is protecting the spring. Secondly, it provided clean water as the cattle and other livestock didn't have access to water. The oozing water was taken in a channel to the lands below for irrigation.

Water lifting devices

Not much work has been done on water lifting devices. For a fairly large farm pond, diesel operated pump remains the only economical gadget to lift water. In small sized farm ponds, local water lifting devices such as moth or dhekli are used. In arid areas, solar operated, pumps may be preferred in years to come. A solar pump with a panel area of 1 m^2 photo voltaic system can approximately irrigate 1 ha during *kharif* (rainy) and *rabi* (post rainy) seasons and 0.3 to 0.4 ha during summer where the water table is available at a depth of 10 m.

Evaporation losses

The water that is lost to the air from the surface of the pond is called evaporation. The amount of water lost by evaporation depends largely on local climatic conditions. Evaporation will also depend on the water surface area. The larger the pond, the more water will evaporate from its surface.

High air temperatures, low humidity, strong winds and sunshine will increase evaporation. Low air temperatures, high humidity, rainfall and cloud cover will decrease evaporation. Evaporation rates are usually expressed as the water depth lost in millimetres over a period of time, *e.g.*, 2 mm day^{-1}, 14 mm $week^{-1}$ or 60 mm $month^{-1}$.

Methods to reduce evaporation

The methods generally used or being tried are broadly listed below:

1. Windbreakers
2. Covering the water surface
3. Reduction of exposed water surface
4. Underground storage of water
5. Integrated operation of reservoirs
6. Treatment with chemical Water Evapo Retardants (WER).

Floating material *viz;* polyethylene sheet, polystyrene sheet, bamboo and *Saccharum munja* covering the water surface reduces the area where evaporation can occur. Evaporation reduction during test period ranged from 37% for *Saccharum munja* to 82% for polystyrene sheet. Foamed rubber sheet, polyethylene sheet and bamboo reduced evaporation by 74%, 66% and 53%, respectively. Rubber and plastic floats can cut evaporation losses by 80%. White polyethylene sheets used for shading the water surface reduce the evaporation by 91%.

Performance

- Runoff water collected in the farm pond could be recycled for protective irrigation at least two times to a depth of 4–5 cm to protect one third of the catchment cropped area during the long dry spells to obtain normal grain yield.
- The lining material used in the farm pond minimizes the seepage losses to 62 litre m^{-2} day^{-1} from 137 litre m^{-1} day^{-1} as in the case of unlined.
- The fish culture in dry land areas (when rainfall is >650 mm) is technically feasible and economically viable and could be adopted by marginal and small farmers to obtain Rs. 2000–3000 monetary benefits in 5 to 6 months. The fish breeds *viz.* Common carp, Catla, Rohu and Grass carp in the ratio of 4:3:2:1 performed better in their growth and development.

Watershed department has taken up excavation of farm ponds. Further, if these farm ponds are lined with this technology seepage losses could be minimized and stored water could be used for multipurpose.

Table 74: Efficient utilization of farm pond water for double crop production under dryland condition

Treatments	Green forage yield (t ha^{-1})	Dry chilli yield (kg ha^{-1})	French bean yield (kg ha^{-1})
S.A. Maize	21.6	924.3	5460
Sweet Sorghum	13.4	832.8	6940
Giant Bajra	28.2	762.8	6640
C.D. (p=0.05)	1.55	91.3	1143

(Anon., 2008)

Harvested rain water in farm pond can be utilized for double cropping system *viz.,* Fodder crops-transplanted Chilli/French bean was found to be more profitable for dry land condition.

2. Check dams

In the watershed area, construction of check dams at the strategic points helps in storing runoff water from the upstream side which could be of great help in ground water recharge and multipurpose uses. Check dams are constructed of rock, sediment retention fiber rolls, gravel, sandbags, or other proprietary product placed across a natural or manmade channel or drainage ditch. A properly designed, constructed, and maintained check dam will reduce scour and channel erosion by reducing flow velocity and encouraging sedimentation.

Appropriate Applications

A check dam either filters the water for sediment as it passes through the dam or retains the water, allowing the sediment to settle while the water flows over the dam. Check dams may be installed:

- In small open channels.
- During the establishment of grass linings in drainage ditches or channels.
- In temporary ditches where the short length of service does not warrant establishment of erosion-resistant linings.

Types

1. **Rock Check dams** are constructed primarily of riprap and are more effective in ditches where the velocity of runoff is expected to be high, or in situations where the surface area exceeds 5 acres and drainage is funneled into a ditch or channel. An erosion control geotextile should be used at the bottom or base of the rock check dam to prevent undercutting, In some instances, erosion control geotextile should also be installed on the overflow portion of the dam to prevent erosion.
2. **Sediment Retention Fiber Roll Check dams** are made from natural or organic material (*e.g.*, straw, coconut fiber (coir), wood fiber (excelsior) mulch, or compost) wrapped in biodegradable netting. Sediment retention fiber roll can be cut to length and used for ditch or channel protection and are easily installed. A series of stair steps made of fiber rolls is effective in filtering and collecting water. Most fiber rolls are degradable and can be left in place after construction.
3. **Sandbag Check dams** are best suited for emergencies such as floods or slides when other materials may be difficult to acquire. Individual sand bags can be filled and placed to form a check dam where water needs to be diverted or channeled. Tightly filled the bags with and stack them using a pyramid approach. Place the bags in an interlocking pattern to assure proper sealing and stability. The upper rows of sandbags shall overlap the joints in lower rows. Bags shall not be stacked higher than 3 feet.
4. **Straw Bale Check dams** should only be used in emergencies and are better suited as a temporary berm or dike to capture or direct runoff where structural strength is not required.

Limitations of Check dams

- Shall be used on slopes with a gradient of 3:1 or less.
- Shall not be used in live streams, except as allowed with proper approval and permitting.
- Shall not be placed in channels that are already grass-lined unless erosion is expected, as installation may damage vegetation.
- Shall not be constructed from silt fence.

- Require extensive maintenance following high velocity flows.
- Promote sediment trapping, which can be re-suspended during subsequent storms or removal of the check dam if not properly maintained.
- Shall be constructed of straw bales only in emergencies.

3. Nala bunds

In the watershed areas differing in size nala bunds constructed can cause for improving the ground water hydrology. Nala bunds are embankments constructed across nalas for checking velocity of runoff, increasing water percolation and improving soil moisture regime. In fact nala bunds and percolation tanks both terms are synonymous, but still they are being called differently at different places. However, the structures costing ₹5 lakh or more are mostly called as percolation tanks. Main objectives of nala bunding are to impound surface runoff coming from the catchments and stabilise the nala grade to facilitate percolation of stored water into the soil sub-stratra with a view to raise ground water level in the zone of influence of the nala bund and to hold the silt flow which would otherwise reach the multipurpose reservoirs and reduce their carrying capacity.

Site Characteristic and Design Guidelines

For selecting a site for Check dams/Nala bunds, the following conditions may be observed

(a) The total catchment of the nala should normally be between 40–100 hectares though the local situations can be guiding factor in this.

(b) The rainfall in the catchment should be less than 1000 mm $annum^{-1}$.

(c) The width of nala bed should be at least 5 meters and not exceed 15 metres and the depth should not be less than 1 metre.

(d) The soil downstream of the bund should not be prone to water logging and should have pH between 6.5–8.

(e) The land downstream of check dam or bund should have irrigable land under well irrigation (This is desirable but not an essential requirement)

(f) The Nalabunds should be preferably located in area where contour or graded bunding of lands have been carried out.

(g) The rock strata exposed in the ponded area should be adequately permeable to cause ground water recharge through ponded water.

(h) Nala bund is generally a small earthen dam with a cutoff core wall of brick work, though masonry and concrete bunds or plugs are now prevalent.

(i) For the foundation of core wall, trench is dug 0.6 m wide in hard rock or 1.2 m in soft rock of impervious nature. A core brick cement wall is created 0.6 m wide to stand atleast 2.5 metres above nala bed and the remaining portion of trench is back filled on upstream side by impervious clay. The core wall is buttressed on both sides by a bund made up of local clays and on the upstream face, stone pitching is done.

(j) Normally the final dimensions of the Nala bunds are: length 10–15 metres, height 2–3 metres and width 1–3 metres, generally constructed in a trapezoidal form. If the bedrock is highly fractured, cement grouting is done to make the foundation leakage free.

(k) Dams should be built at sites that can produce a relatively high depth to surface areas so as to minimize evaporation losses.

(l) Rocky surfaces should not he fractured or cracked, which may cause the water to leak away to deeper zones or beneath the dam.

(m) Dam foundation must be of solid impermeable rock with no soil pockets of fracture lines.

(n) Convenient location for user groups.

(o) No soil erosion in the catchment area.

(p) Dams should be sited along the edges of depressions or directly across the lower ends of deep gullies into the rock.

4. Roof water harvesting

In the urban areas and also industrial establishments water is not only becoming costly but also scarce. Latest assessments on water availability in the globe delineated many cities as dry areas in future if adequate water harvesting, storing and judicious using methods are not followed and situation is going to be grim. Hence, it is the responsibility of owners of every house and industries to harvest rain water and use it efficiently by adopting rooftop rain water harvesting technique. In the urban areas it should be made mandatory. Rooftop rain water harvesting is the technique through which rainwater is captured from the roof catchments and stored in reservoirs. Harvested rainwater can be stored in sub surface ground water reservoir by adopting artificial recharge techniques to meet the household needs through storage in tanks. Capturing and storing rain water for use is particularly important in dryland, hilly, urban and coastal areas.

The Main objective of rooftop rainwater harvesting is to make water available for future use.

Need for Rooftop rainwater harvesting

1. To meet the ever increasing demand for water
2. To reduce the runoff which chokes storm drains
3. To avoid flooding of roads

4. To augment the groundwater storage and control decline of water levels
5. To reduce groundwater pollution
6. To improve the quality of groundwater
7. To reduce the soil erosion
8. To supplement domestic water requirement during summer, drought *etc.*

Advantages

1. Harvested water can also be used for household purpose.
2. It can also be used for nourishing the ornamental plants and kitchen garden.
3. Provides self–sufficiency in water supply
4. Reduces the cost for pumping of groundwater
5. Provides high quality water, soft and low in minerals
6. Improves the quality of ground water through dilution when recharged the ground water
7. Reduces soil erosion in urban areas
8. The rooftop rainwater harvesting is less expensive
9. Rainwater harvesting systems are simple which can be adopted by individuals
10. Rooftop rain water harvesting systems are easy to construct, operate and maintain

Disadvantages

1. The success of rainfall harvesting depends upon the frequency and amount of rainfall: therefore, it is not a dependable water source in times of dry weather or prolonged drought.
2. Low storage capacities will limit rainwater harvesting so that the system may not be able to provide water in a low rainfall period. Increased storage capacities add to construction and operating costs and may make the technology economically unfeasible, unless it is subsidized by government.
3. Leakage from cisterns can cause the deterioration of load bearing slopes.
4. Cisterns and storage tanks can be unsafe for small children if proper access protection is not provided.
5. Possible contamination of water may result from animal wastes and vegetable matter.
6. Where treatment of the water prior to potable use is infrequent, due to a lack of adequate resources or knowledge, health risks may result; further, cisterns can be a breeding ground for mosquitoes.

7. Rainfall harvesting systems increase construction costs.
8. Rainfall harvesting system may reduce revenues to public utilities.

Formula

Total quantity of water to be collected (cu.m.) = Roof top area (Sq.m.) × Average monsoon rainfall (mm) × 0.8

E.g: Rainwater harvesting potential in urban areas is huge. Considering the availability of rainwater in a residential site of 40 feet × 60 feet (an area of 2400 sq.ft./223 sq.mts.), around 1,78,400 litre of rainwater can be harvested in a location where the rainfall is around 1000 mm. The amount of rainwater that can be harvested from the available rainwater in the plot depends on potential rainfall, catchment area available, collection methods and its efficiency *etc.*

Roof water harvesting

Chapter 9

Crops and Cropping Systems for Dryland Agriculture

The occurrence of periods of water deficit for crop production, referred to as climatic drought, is commonly observed and leads to low water availability for crops. Besides climatic drought, crop water stress may also result from low levels of plant available water in the soil profile due either to the existence of physical barriers to water infiltration (*e.g.*, surface sealing) or to soil chemical or physical limitations to plant root growth and root water uptake. Drought resulting from such factors will be referred to as edaphic drought since it is caused by soil specific conditions rather than by limited rainwater supply, and can occur even under conditions of sufficient and well-distributed rainfall. Finally, even where water is very scarce, particularly in the driest areas, a surprisingly small proportion of the available water is actually transpired by the crop. Non-productive losses include surface runoff, deep drainage, evaporation from the soil surface and deep cracks, and transpiration by weeds. Dry land farming is a system of crop cultivation depending on rainfall as the only source of moisture and the technologies refers to the cropping pattern/rotation and agronomic practices adjusted and adopted in accordance with the moisture regime. The crops selected should have short duration, deep root system, erect leaves, moderate tillering and photosynthetically efficient.

Plant Ideotypes: The term ideotype was introduced by Donald (1968). According to him ideotype is a biological model, which is expected to perform or behave in a predictable manner within a defined environment. This term has synonyms *viz.,* Model plant type, Ideal plant type.

In case of cereals, a crop ideotype is erect, sparsely-tillered plant, with small erect leaves.

Main traits of the ideotype should be: Satisfactory levels of production (both quantitative and qualitative) a deep root system, being able to establish symbiotic relationships with soil micro-organisms. Moreover it should compete with weeds.

Factors affecting ideotypes

- **Crop species:** In monocots, tillering is more important but in dicots branching is important.

- **Cultivation:** The features of irrigated crops differ from that of rainfed crop.
- **Socio-economic condition of farmers:** *e.g.*, dwarf crop is ideal for mechanical harvesting.
- Economic use.

Ideotypes for dryland areas
- **Crop duration:** short duration is essential for dryland areas as it matures early.
- **Plant architecture:** Short stature, moderate tillering, sturdy stems and dark green leaves. Elastic plant (capacity to adjust its volume with variation in soil moisture) is better than non-elastic plant.
- **Intra-plant competition**: should be less and has the capacity to remobilize photosynthates to the economic part at the end season drought.
- **Drought tolerance:** should have deep root system with maximum branching at deeper zones, drought tolerant mechanisms (Leaf drooping, shedding leaf/ branches, chemical balance in stomata *etc.*)
- **Pest and diseases resistance**
- **Input/Nutrient responsive**

Wheat ideotype
- A short strong stem
- Erect leaves
- Few small leaves
- Larger and an erect ear
- Presence of awns

Rice ideotype
- Erect, short and thick leaves
- Semi-dwarf stature
- High tillering capacity
- More panicles
- High harvest index

Maize ideotype
- Small tassel size
- no tillers
- Large cobs
- Angled leaves for good light interception

Cotton ideotype

- Short stature
- Medium to big boll size (3.5 to 4 g)
- High degree of resistance to insects diseases
- Few smaller & thick leaves with sparse hairiness

Crop planning

Crop planning is based on rainfall amount and its distribution, length of growing season and moisture availability period and water demand of crops/cropping system.

Based on rainfall, soil characters, period of moisture availability and meteorological parameters, different suitable cropping systems have been identified for various climatic regions of the country (Table 75).

Table 75: Suitable cropping systems based on rainfall and water availability period

Rainfall (mm)	Soils	Water availability (Weeks)	Potential cropping system
350–600	Alfisols & shallow Vertisols	20	Single *Kharif* cropping
350–600	Aridisols & Entisols	20	Single cropping either in *Kharif* or *Rabi*
350–600	Deep Vertisols	20	Single *Rabi* cropping
600–750	Alfisols, Vertisols & Entisols	20–30	Intercropping
750–900	Entisols, Deep Vertisols, Alfisols & Inceptisols	>30	Double cropping with monitoring
>900	Entisols, Deep Vertisols, Alfisols & Inceptisols	>30	Assured double cropping

Different crops have been identified based on Length of growing period (Table 76). Length of growing period is the period in days during a year or season when precipitation (P) exceeds half of the Potential Evapotranspiration (PET) during vegetative phase till soil moisture storage meets 25% of PET during maturity or harvesting stage of the crop.

Table 76: Length of Growing Period-choice of cropping system

Length of Growing Period	Cropping system
< 75 days	Grasses, shrubs, silviculture Monocropping of short duration pulses
75–140 days	Monocropping
140–180 days	Intercropping
180–240 days	Double cropping
> 240 days	Multiple cropping

Crops and cropping systems

Different crops and cropping systems for dry farming regions of India have been identified and evaluated in different agro climatic regions of the country (Table 77). Based on length of crop growing period in different soils, different crops and cropping systems are suggested in different agro climatic regions as follows:

Table 77: Cropping systems in important dry farming regions of India

Region	States	Places	Soils	Crops/cropping systems
Jhansi	Uttarpradesh	Jhansi, Hamirpur, Banda, Lalitpur, Jalaun	Red black	• Sorghum-safflower/mustard • Cowpea/urd/moong, -gram/ safflower • Rice-soybean-gram/ safflower
Rajkot	Gujarat	Rajkot, Surendranagar; Jamnagar, Junagadh Bhavanagar, Amreli	Medium black	• Sorghum/bajra/cotton/green gram/black gram redgram/ cluster bean/groundnut/ sesamum/castor-safflower/ sunflower/green gram/ mustard.
Akola	Maharashtra	Akola, Amravati, Wardha, Yeotmal, Parbhani, Buldana, Khandesh, Adilabad,	Medium and deep black	• Green gram/sorghum/ sunflower/cotton + green gram/groundnut-sorghum • Groundnut + sunflower
Sholapur	Maharashtra	Solapur, Ahmednagar, Nasik, Pune, Satara, Sangli, Dhule, Bhir, Osmanabad, Jalgaon, Buldhana	Black	• Pear millet-Gram • Sorghum/Pearl millet + Redgram - Horsegram • Redgram + setaria/ Groundnut/sunflower • Castor-Horsegram
Indore	Madhya Pradesh	Indore, Ratlan Ujjain Dewar, Dhar, Khargaon Khandura,	Medium deep black	• Maize-gram/safflower • Sorghum + soybean-gram/ safflower • Maize + groundnut • Sorghum + redgram

Region	States	Places	Soils	Crops/cropping systems
Rewa	Madhya Pradesh	Sidhi, Rewa, Satna, Shadol, Panna, Jabalpur, Damoh, Chattarpur, Tikamgarh	Medium black mixed red and black	• Sorghum +Redgram-gram • Rice- wheat/gram • Black grim Green gram-wheat • Rice-lentil
Bijapur	Karnataka	Bijapur, Gulbarga, Belgraum, Raichur	Medium and deep black	• Green gram-sorghum/ safflower • Groundnut/pearl millet + redgram. Bengal gram+ safflower/cotton
Udaipur	Rajasthan	Udaipur, Chittorgarh, Bhilwara, Ajmer, Banswara, Dungarpur	Medium black	• Sorghum maize-safflower mustard • Pearl millet + cowpea - mustard • Sorghum-mustard • Redgram/green gram/ groundnut/Sunflower-wheat/ mustard
Bellary	Karnataka	Chellakere, Chitradurga, Bellary, Raichur	Medium and deep black	• Sorghum/safflower/gram • Sorghum + lablab
Agra	Uttar Pradesh	Agra, Aligarh, Mathura, Etah, Manipuri	Deep alluvial sandy loam	• Pearl millet/black gram/ green gram/redgram/cluster bean/groundnut safflower/ mustard/ • Pearl millet+ redgram/black gram/greengram/groundnut + castor
Anantapur	Andhra	Anantapur, Karnool, Chithoor	Red loam	• Pearl millet/sorghum/setaria/ castor/Redgram/groundnut/ mesta • Groundnut + Redgram/ castor/pearl millet

Selection of suitable crops and cropping systems

The compatible cropping system under dryland condition should possess high degree of drought tolerance, elasticity in growing period based on available farm resources, high fertilizer responsive (kg grain per kg of fertilizer application), suitable for intercropping systems, photo insensitive, higher root/ shoot ratio *etc.*

Sowing window

In fingermillet sowing window starts from July 15 and continues upto September 1st week depending upon the onset and distribution of rainfall (Table 78) in south eastern dry zone of Karnataka. Another advantage is availability of long, medium and short duration varieties in finger millet. It is advantageous to fit long duration varieties during July, medium duration varieties during August and short duration varieties during August second fortnight to first week of September depending upon the rainfall. So in finger millet selection of variety for the sowing window is a real time natural contingency measure.

Table 78: Sowing window for Finger millet under dryland conditions

Variety	Sowing time	Duration (Days)
Long duration		
MR-1	June–July	120–125
MR-6	June–July	120–125
Medium duration		
GPU–28	July–August	110–115
GPU–66	July–August	110–115
ML-365	June–August	105–110
Short duration		
GPU–48	August	100–105
GPU–45	All season	95–100
KMR-204	July–August	100–105

Table 79: Transplanting versus direct sown fingermillet

Treatments	DOS/ DOT	Yield (kg ha^{-1})		% increase in yield	Net returns (₹ha^{-1})	B:C ratio	RWUE (kg ha-mm^{-1})
		Grain	Straw				
Transplanted (MR-1)	23rd Aug	2667	4100	7	35742	2.37	5.24
Direct sown (MR-1)	24th July	2500	4370	–	33072	2.31	4.39

Source: Ramachandrappa *et al.*, 2019

Another real time contingency measure in fingermillet is the technique of transplanting which can be adopted by raising seedlings wherever there is availability of water in the field or backyard with right variety of fingermillet and using the seedlings for transplanting after the receipt of rains (Table 79).

Other best suited crops for rainfed conditions include Bajra, Foxtail millet, Little millet, Proso millet, Barnyard millet and Kodo millet. These millets are called famine reserves since, they can be stored without storage pests for several years. They are short duration and drought tolerant. They are rich in dietary fibres and very well preferred for diabetic patients and those who are suffering from heart diseases. With the increased awareness about these millets from among the calorie conscious city elites, there is increased demand for these minor millets grown under rainfed conditions. There is a urgent need for their seed production and promotion of these minor millets in the rainfed areas. Bajra has the dual advantage of not only drought tolerant but also heat tolerant and well suited for dryland areas. The details of these millets inter alia varieties and inputs for their production is given below:

Bajra

Foxtail millet

Kodomillet

Littlemillet

Prosomillet

Barnyard millet

Pearlmillet

Varieties	Sowing season	Duration (Days)	Spacing	FYM t ha^{-1}	Fertilizers (kg ha^{-1})	Seed rate (kg ha^{-1})	Yield (Q ha^{-1})
WCC-75	January, June – July	85–90	45 × 10–15	6.0	100:65:25	15	12–14
PHVB-910	,,	85–90	,,	,,	,,	,,	12–14

Foxtail millet

Varieties	Sowing season	Duration (Days)	Spacing	FYM t ha^{-1}	Fertiliz-ers (kg ha^{-1})	Seed rate (kg ha^{-1})	Yield (q ha^{-1})
PS-4	Monsoon May-2nd week	85–90	25–30 × 10 cm	5.0	40:20:0	10–15	18–20
TNAU–186		85–90	25–30 × 10 cm	5.0	40:20:0	10–15	18–20
Nasimharaya		85–95	25–30 × 10 cm	5.0	40:20:0	10–15	20–22
Krishnadevaraya		85–96	25–30 × 10 cm	5.0	40:20:0	10–15	20–22
Shree lakshmi		75–80	25–30 × 10 cm	5.0	40:20:0	10–15	20–23
TSAU–196		85–90	25–30 × 10 cm	5.0	40:20:0	10–15	18–20

Little millet

Varieties	Sowing season	Duration (Days)	Spacing	FYM t ha^{-1}	Fertilizers (kg ha^{-1})	Seed rate (kg ha^{-1})	Yield (q ha^{-1})
OLM-203	Monsoon	75–85	25–30 × 10 cm	5.0	20:20	10–12	10–12
Q-2	Monsoon	70–80	25–30 × 10 cm	5.0	20:20	10–12	10–12
K.K.8	Monsoon	70–80	25–30 × 10 cm	5.0	20:20	10–12	10–13

Proso millet

Varieties	Sowing season	Duration (Days)	Spacing	FYM t ha^{-1}	Fertilizers (kg ha^{-1})	Seed rate (kg ha^{-1})	Yield (q ha^{-1})
K–1	Monsoon	70–75	25–30 × 10 cm	5.0	20:20:0	10–15 kg	9–10
GPU–8	Monsoon	70–75					15–16
GPU–21	Monsoon	67–72					15–18

Barnyard millet

Varieties	Sowing season	Duration (Days)	Spacing	FYM t ha^{-1}	Fertilizers (kg ha^{-1})	Seed rate (kg ha^{-1})	Yield (q ha^{-1})
V. L. – 29	Monsoon	85–90	25–30 × 10 cm	5.0	20:20:0	10/15	15–18
V. L. – 172	Monsoon	90–95	25–30 × 10 cm	5.0	20:20:0	10/15	20–22

Kodomillet

Varieties	Sowing season	Duration (Days)	Spacing	FYM t ha^{-1}	Fertilizers (kg ha^{-1})	Seed rate (kg ha^{-1})	Yield (q ha^{-1})
GPUK–3	June–July	110–105	25–30 × 10 cm	5.0	30:15:0	10/15	18–20
BK–155	June–July	105– 110	25–30 × 10 cm	5.0	30:15:0	10/15	20–22

Red soil areas

Improved varieties

Improved high yielding varieties of pigeon pea, field bean and vegetable cowpea recommended for rainfed situations of southern dry region of Karnataka based on their long term performance in research stations and farmers field (Table 80).

Table 80: Improved high yielding varieties of pulses for rainfed situations

Crops	Variety	Duration (days)	Yield (kg ha^{-1})	Special characters
Pigeonpea	TTB-7	150–200	1200–1500	Suitable for dhal and intercropping system in early *kharif*
	BRG-1	170–190	1200–1500	Suitable for dhal, vegetable and intercropping system in early *kharif*
	BRG-2	150–170	1250–1600	Suitable for dhal, vegetable and delayed sowing
Vegetable cowpea	PKB-4	80–85	1000–1200	Suitable for double cropping in early *kharif*
	PKB-6	80–85	1000–1200	Suitable for double cropping in late *kharif*
Field bean	HA-3	95–105	1000–1250	Short duration, photo-insensitive suitable for inter/double cropping
	HA-4			

Source: Ramachandrappa *et al.* (2014)

Pigeonpea varieties *viz.,* TTB-7, BRG-1 and BRG-2 are suitable for sowing in May (early *kharif*) and in intercropping systems with groundnut and finger millet. PKB-4 and PKB-6 are short duration vegetable cowpea varieties (80–85 days) and well suited for double cropping. Field bean variety HA-3 and HA-4 are suitable for intercropping in pigeonpea because of its short duration, photo-insensitivity and good fragrance. The horse gram variety PHG-9 is high yielding, fairly tolerant to powdery mildew and having bold grain.

The improved high yielding varieties of groundnut, sunflower, sesame, niger and castor are recommeneded for rainfed situations of *Alfisols* (Table 81).

Table 81: Improved high yielding varieties of oilseeds for rainfed situations

Crops	Variety	Duration (days)	Yield (kg/ha)	Special characters
Groundnut	TMV–2	100–120	800–1000	Suitable for crop rotation and intercropping
	JL-24	90–95	750–950	
	Chintamani-2	100–120	800–1000	

Sunflower	KBSH-44	90–95	1500	Suitable for delayed onset of monsoon as a contingency crop
	KBSH-53	90–95	1500	
Sesame	TMV-3	85–90	400–500	Suitable for double cropping
Niger	KBN-1	80–90	400–500	
Castor	DCS-9 (Jyothi)	140–150	1000–1200	Suitable for intercropping in groundnut

Source: Ramachandrappa *et al.* (2014)

Crops for different times of sowing in black soils

June: Hybrid bajra, pigeon pea, spreading groundnut in shallow soils, jowar, hybrid bajra, groundnut, pigeon pea, green gram and black gram in medium deep black soils.

July 15–30: Hybrid bajra, pigeon pea, spreading groundnut, sunflower, setaria, sesamum, niger, castor, horsegram in shallow soils; hybrid bajra, spreading groundnut, pigeon pea, sunflower in medium deep black soils.

August: Sunflower, castor, sesamum, niger, horsegram in shallow soils; cotton, sunflower in medium and deep soils.

September 15–30: Sunflower in shallow soils; sunflower, safflower, *rabi jowar* in medium deep soils.

October upto 15 November: Safflower, *rabi* jowar, wheat, gram in medium and deep soils.

Upto end of October: Wheat, bengal gram in deep black soils.

Intercropping

In the red soil areas of southern districts of Karnataka, fingermillet is a predominant millet crop grown under rainfed conditions. Since intercropping is an insurance against total crop failure, intercropping of pigeonpea in fingermillet in the ratio of 8:2 (Table 82) with a conservation furrow in between two rows of pigeonpea is also drought mitigation strategy under water scarcity conditions, conservation furrow acting as a sink for rain water collection and also a channel for safe disposal of excess rainfall during extreme situations. This popular intercropping system would help in achieving not only higher Finger millet equivalent yield but also B:C ratio (Table 83). It can also address food and nutritional security issues by supplying low glycemic index millet and pulse apart from adding 2–3 tonnes of foliage droppings and fixing of atmospheric nitrogen and enriching the soil fertility. After harvest pigeonpea stalks are being used as fuel or thatching purpose in the rural household.

Table 82: Intercropping systems recommended with sowing window

Sowing time	System	Ratio of rows
May, June	Redgram + Maize	1:1
May, June	Redgram + Short duration cowpea	1:2
June, I fortnight of July	Fingermillet + Redgram	8:2
June, I fortnight of July	Groundnut + Redgram	8:2
June	Groundnut + Castor	8: 1
June, July	Finger millet + Field bean	8: 1/1 0:1
June, July fortnight of August	Fingermillet + Soybean	1:1/8:1
May June	Jowar + Redgram	2: 1
May	Sesamum + Redgram	10:2 & 3:1
May, June	Castor + Niger/Finger millet	3: 1
August, September	Horsegram + Niger	8:2

Cereal based Intercropping

Maize based inter cropping system

Maize + Soybean (2:1)

Maize + Pigeon pea (1:1)

Sorghum based inter cropping system

Sorghum + Pigeon pea (2:2)

Sorghum + Groundnut (2:4)

Finger millet

Finger millet + Pigeon pea (8:2)

Finger millet + Field bean (8:1)

Finger millet + Soybean (1:1)

Table 83: Finger millet + Pigeon pea (8:2)

Treatments	Yield (q ha^{-1})		FMEY (q ha^{-1})	Straw Yield (q ha^{-1})	Gross returns (₹ ha^{-1})	Net Returns (₹ ha^{-1})	B:C Ratio
	FM	PP					
Finger millet + Pigeon pea (8:2) with conservation furrow	26.67	4.17	38.61	54.10	46534	31534	3.10
Farmer's Practice	18.94	–	18.94	59.40	25289	11989	1.90

Pulse based Intercropping systems

The pigeonpea + Field bean (1:1) intercropping system helps in achieving higher pigeonpea quivalent yield, and B:C ratio (Table 84 and 85).

Few examples for pulse based cropping systems are:

Pigeonpea + Soybean/Cowpea/Black gram/Green gram(1:1)

Sesamum + Pigeonpea (8–10: 2 or 3)

Field bean + Castor (4–6:1)

Table 84: Pulse based intercropping system

Treatments	Yield (kg ha^{-1})		PP equivalent yield (kg ha^{-1})	Returns (₹ ha^{-1})		B:C Ratio
	PP	Intercrop		Gross	Net	
PP + Field Bean (1:1)	842	416	1224	41965	29165	3.27
Farmer's Practice	453	–	453	15855	3455	1.28
PP + Cowpea (1:1)	1212	200	1383	48405	35805	3.84
Farmer's Practice	476	–	476	16660	4260	1.34

Table 85: Yield and economics of pulse based production system

Treatments	Yield (kg ha^{-1})		Stalk/haulm yield (kg ha^{-1})	Gross income (₹ ha^{-1})	Net Income (₹ ha^{-1})	B:C ratio
	Pigeonpea	Filed bean				
Pigeonpea + Fieldbean (1:1) (TTB-7 + HA–4)	970	460	Pigeonpea stalk 3880 Fieldbean haulm 3680	51530	39030	4.12
Pigeonpea sole crop (TTB-7)	1050		4200	38850	26500	3.15
Fieldbean sole crop (HA-4)		980	7840	33320	22070	2.96

Pigeonpea is a major pulse crop occupying the field for 180–200 days. Intercropping in pigeonpea has the following advantages:

1. Pigeonpea is slow in its growth during early stages which can be exploited for intercropping short duration crops.
2. Pigeonpea is widely spaced at 60–90 cm between rows, this wider corridor facilitates fitting a suitable intercrop.
3. Pigeonpea invariably requires two sprays of plant protection chemicals to control pod borer, the expenditure for which can be obtained from short duration intercrops.
4. Intercropping helps in efficient use of space, moisture, nutrients and solar energy.
5. Intercrops like cowpea or fieldbean can be harvested either for green pods or for grains depending upon the rainfall conditions.

6. After harvesting intercrops plant residue can be used as a fodder during scarcity situations to sustain the livestock.
7. Residues of intercrops can also be used as mulch between two rows of pigeonpea for the conservation of moisture and also to reduce the impact of raindrops.
8. A compatible intercrops like cowpea and fieldbean in pigeonpea can effectively smother weeds.
9. It is possible to achieve higher pigeonpea equivalent yield and B:C ratio.

Oilseed based Intercropping

Groundnut is one of the major oilseed crops grown under rainfed conditions. Groundnut + Castor (8:1) and Groundnut + Pigeonpea (8:2) can be grown as suitable intercrops in groundnut. After the harvesting of groundnut, castor and pigeonpea can grow rapidly with more number of branches covering the wider corridor left. Thus facilitating easy movement for plant protection sprays and harvesting apart from achieving higher groundnut equivalent yield and B:C ratio.

Table 86: Yield and economics of Groundnut + Castor (8:1) cropping system

Treatments	Yield (kg ha^{-1})		Pod equivalent yield (kg ha^{-1})	Returns (₹ ha^{-1})		B:C Ratio
	GN	Castor		Gross	Net	
Groundnut + Castor (8:1)	1456	62	1543	38575	21975	2.32
Farmer's practice	728	52	801	20025	3425	1.20

Table 87: Yield and economics of groundnut based crop production system

Treatments	Yield (kg ha^{-1})					Gross Income (₹ ha^{-1})	Net Income (₹ ha^{-1})	B:C ratio
	Groundnut		Pigeonpea	Castor	*Akkadi* crops (kg ha^{-1})			
	Pod	Haulm						
Groundnut + Pigeon pea (8:2) (JL-24 + TTB-7)	790	3190	645			46290	29290	2.72
Groundnut + Castor (8:1) (JL-24 + DCS-9)	885	3540		440		36230	19730	2.20
Groundnut + *Akkadi* *(Farmer's practice)*	650	2600			Pigeonpea–110 Field bean-80 Sorghum-30, Castor-60	27550	13550	1.97

Source: AICRPDA

Inter cropping systems for Vertisols

Shallow black soil (*Kharif*)

Hy. Bajra + Redgram (2:1)

Setaria + Redgram (4:2)

Medium deep black soil (*Kharif*)

Sunflower + Redgram (2:1)

Groundnut + Redgram (4:2)

Redgram + Blackgram/Greengram (1:1)

Cotton + Groundnut (1:2)

Groundnut + Hy. Bajra (4:2)

Hy. Bajra + Redgram (2:1)

Hy. Sorghum + Redgram (2:1)

Blackgram/greengram+Redgram(4:2)

Sesamum + Redgram (4:2)

Chilli + Cotton (1:1)

Deep black soil (*Rabi*)

Safflower + Bengalgram (2:4)

Rabi Jowar + Bengalgram (2:1)

Coriander + Safflower (4:2)

Maize + Redgram (2:1)

Double cropping systems

Alfisols

Cowpea (May) – ragi (August), sesamum (May) – horsegram (September)/ Transplanted ragi, groundnut (May) – Transplanted ragi (September), jowar (April) – ragi/horsegram (Table 88).

Table 88: Double cropping systems recommended

April, May	Sesamum-Finger millet Fodder Crop-Finger millet
May, 1st fortnight of June	Fodder crop- Chilli Cowpea–Finger millet
May	Mungbean- Finger millet
April, May	Jowar- horsegram
June	Jowar- Transplanted finger millet. Bajra– Horsegram Sunflower- Jowar

Second crop to be sown soon after the harvest of first crop

Economics of Double cropping system

Livestock is an important component and a subsidiary enterprise under rainfed conditions. Fodder production should be given equal importance for the sustainable livestock production. In the Alfisol situations of southern Karnataka with bimodal distribution of rainfall, early pre-monsoon rains in the month of April/May can be effectively be used for raising fodder crops *viz.,* fodder sorghum, fodder maize and fodder bajra. Water collected in the farm pond can be productively be used during dryspells for these fodder crops through sprinkler irrigation. After the harvest of the green fodder either for direct feeding or for silage purpose, second crop of chilli, tomato, beans, onion, chickpea *etc.*, can be successfully raised with one or two protective irrigation from the collected water in the farm pond. Data in the table indicated superiority of giant bajra followed by chickpea in achieving fodder security apart from higher RWUE, system net returns and B:C ratio.

Table 89: Performance of forage crops and chickpea under double cropping system:

Treatments	Yield (kg ha^{-1})		RWUE (kg ha mm^{-1})	
	Crop1	Crop 2	Crop 1	Crop 2
Main plot				
South African maize (SA Tall)	37980	775	216	
Sweet sorghum (SSV-74)	34765	884	198	
Giant bajra	43067	1067	245	
S.Em ±	873	23.14		
C.D (P = 0.05)	3427	90.88		
Sub plot				
Annigeri-1		868		
JG-11		950		
S.Em ±		20.63		
C.D (P = 0.05)		71.41		
Sub-Sub plots				
100% RDF		954		
75% RDF		863		
S.Em ±		21.29		
C.D (P = 0.05)		65.63		

Source: Anon., 2019

Table 90: Double cropping using pond water

Treatment	Gross Returns (₹ ha^{-1})		System Gross Return (₹ ha^{-1})	Net Returns (₹ ha^{-1})		System Net Returns (₹ ha^{-1})	B:C ratio		System B:C ratio
	Crop 1	Crop 2		Crop 1	Crop 2		Crop 1	Crop 2	
Main plots									
SA maize (SA Tall)	56970	28692	85662	35555	9767	45322	2.66	1.52	2.12
Sweet sorghum (SSV-74)	52148	32693	84841	33676	13768	47444	2.82	1.73	2.27
Giant bajra	64600	39484	104084	43609	20559	64168	3.08	2.09	2.61
Sub plots									
Annigeri-1		32104	32104		13179	13179		1.70	1.70
JG-11		35142	35142		16217	16217		1.86	1.86
Sub-Sub plots									
100% RDF		35303	35303		13598	13598		1.63	1.63
75% RDF		31943	31943		10244	10244		1.47	1.47

Source : Anon., 2019

Vertisols

Double cropping systems: Sunflower (*Kharif*)–safflower (*Rabi*), bunch groundnut (*Kharif*)–*Rabi* jowar, green gram/black gram (*Kharif*)–*Rabi* jowar/safflower, setaria–cotton.

Crop diversification: Crop diversification can work as drought mitigation strategy under dry land conditions it includes;

- Increasing the area of drought resistant crops
- Increasing the area of castor cultivation
- Increasing the area of small millets (little millet, proso millet, kodo millet, foxtail millet)
- Growing of forest trees or fodder crops on the bunds
- Crop rotation and intercropping system
- Agri-Horti systems
- Agro-forestry
- Soil health improvement through *in-situ* and *ex-situ* green manuring

In rainfed areas, to improve the water holding capacity of the soil and also soil physical, chemical and biological properties, it is essential to apply organic manures in the form of FYM, compost and green manures. Since, the availability of FYM is becoming not only costly but also scarce, hence, green manuring of rainfed soils in the premonsoon season has become a necessity to improve the soil fertility without loosing the regular rainfed crop. Following are some of the green manuring crops with their biomass yield and nitrogen content can be grown depending upon the agro-ecological situation (Table 91).

Table 91: Biomass production potential and N content of green manure crops

Local name	***Green manure (t/ha)***	***Nitrogen (%)***
Sunhemp	21.2	0.43
Dhaincha	20.0	0.43
Pillipsera	18.30	1.10
Mungbean	8.0	0.53
Cowpea	15.0	0.49
Horse gram	28.0	0.33
Guar	20.0	0.34
Senji	28.6	0.57
Khesari	12.3	0.54
Berseem	15.5	0.43
Sesbania	24–26	160–180 kg ha^{-1}
Glyricidia	2–3 loppings	2.5– 2.7
Pongamia	2–3 loppings	3.2–3.5
Water hyacinth	–	3.0–3.2

in-situ Green Manuring for Finger millet

in-situ green manuring of horse gram-sowing finger millet (Cv. GPU–28) + 50% Nitrogen increased grain yield and Straw yield with a B:C ratio of 1.95 compared to 1.74 under farmer's practice.

Horse gram grown during early season

Finger millet (GPU–28) crop raised after Horse gram incorporation

Glyricidia as *ex-situ* green manuring crop

Raising *Glyricidia* plants on the bunds & boundaries (300–325 plants/ha) produced green biomass to supply 25–35 Kg N ha^{-1} besides fixing atmospheric N (30–50 Kg/ha) Green leaf manuring to supply 50% N + 50% NPK in finger millet resulted in higher finger millet (3004 kg ha^{-1}) yield compared to application of Rec. NPK (2572 kg ha^{-1}) besides, maintaining sustained soil fertility.

Glyricidia plants on the bunds

Finger millet + INM

Incorporation of green manure (*glyricidia*) to soil

Chapter 10

Conservation Agriculture

Green Revolution has increased food grain production by four fold since 1950–51 with the adoption of HYVs, intense input use, extensive tillage, burning of residues and irrigation. The intense cultivation has lead to degradation of natural resource such as soil, water, vegetation *etc*. In this context, globally conservation agriculture (CA) has opened a new paradigm as it has potential for higher resource use efficiency, water productivity and climate change mitigation through its key principles. Total area and percentage of total area under no tillage practised as part of conservation agriculture in different continents is given in Table 92.

Table 92: Area under No-tillage by continent

Continent	Area (hectares)	Per cent of total
South America	4,95,79,000	46.8
North America	4,00,74,000	37.8
Australia & New Zealand	1,71,62,000	11.5
Asia	25,30,000	2.3
Europe	11,50,000	1.1
Africa	3,68,000	0.3
World Total	11,08,63,000	100

Source: Derpsch *et al*., 2010

Problems associated with conventional agriculture

- Poorly chosen tillage system
- Reduction of vegetative cover
- Pulverization of surface layer
- Compaction of underlying layer
- Wind erosion
- Reduction of water infiltration
- Bad root development
- Increase in runoff
- Low yields
- High production costs

- Pollution
- Inefficient use of water and fertilizers
- Erosion

The harmful effects of tillage

- Drastically alters original structure
- Breaks up aggregates
- Buries crop residues
- Bares/Exposes soil to the elements
- Reduces biodiversity
- Increase CO_2 emissions
- Creates compact subsoil layers

The reason for poor soil quality and productive capacity of soil are

- Erosion of topsoil and organic matter due to water erosion
- Intensive deep tillage and inversion tillage
    - fast decomposition of crop residues
    - breaking of stable soil aggregates
    - disturbance to the soil flora and fauna
- Mono cropping and low fertilizer application
- Nutrient imbalance due to improper use of primary, secondary and micronutrients
- Poor use of organic manures such as FYM, Compost and no recycling of crop residues due to demand for fodder and fuel
- No or low green manuring as it competes with the regular crop for resources
- Poor nutrient use efficiency attributing to nutrient losses due to leaching, volatilization and denitrification
- Indiscriminate use of agrochemicals resulting in poor soil and water quality
- Water logging, salinity and alkalinity and acid soils

What is Conservation Agriculture?

Conservation agriculture (CA) can be defined by as "a concept for resource–saving agricultural crop production that strives to achieve acceptable profits together with high and sustained production levels while concurrently conserving the environment" (FAO 2007). To conserve the natural resources and to avoid degradation of land and other natural resources, Conservation agriculture has emerged as an alternative strategy for conserving the natural resources.

Globally over the past 2–3 decades, CA has emerged as a way for transition to the sustainability of intensive production systems. The term "Conservation tillage" refers to a number of strategies and techniques for establishing crops in a previous crop's residues, which are purposely left on the soil surface.

Conservation agriculture practices are such practices which include

- Direct sowing
- No-tillage or
- Reduced tillage or minimum tillage
- Surface incorporation of crop residues
- Establishment of cover crops

The principal benefits of conservation tillage are improved rain water conservation and the reduction of soil erosion. Additional benefits include reduced fuel consumption, reduced soil compaction, planting and harvesting flexibility, reduced labour requirements and improved soil health.

The key features of conservation agriculture are

- No ploughing, disking of soil cultivation (*i.e.*, not turning over of the soil)
- Crop and cover crop residues stay on the surface
- No burning of crop residues
- Weedicides for weed management
- Permanent crop and weed residue mulch on soil
- Continuous cropland use
- Crop rotation and cover crops
- Specialized equipment for sowing
- The soil is protected from erosion and runoff
- Less contamination of surface water occurs
- Water retention and storage is enhanced- allow recharging of aquifers
- CA offers a powerful option for meeting future food demands while also contributing to sustainable agriculture and rural development
- CA methods improve the efficiency of input and increase farm income, sustain crop yields, protect & revitalize soil, biodiversity and the natural resource base
- CA methods enhance natural biological processes above and below the ground by reducing interventions such as mechanical soil tillage to an absolute minimum

The principles of conservation agriculture with the most important functions are:

1. **Minimum mechanical soil disturbance**
   - Erosion control
   - Soil carbon build up
2. **Permanent organic soil cover**
   - Erosion control
   - Biodiversity and environment
3. **Diversified crop rotation in the case of annual crops or plant association in case of perennial crops**
   - Pest and disease control
   - Soil structure improvement
   - Biodiversity
4. **Controlling in-field traffic reducing compaction**

   Conservation tillage and residues management influences following soil properties
   - Soil temperature is altered
   - Improves soil aggregation, reduces density and increases porosity
   - Soil organic matter and soil fertility is improved
   - Influence crusting, increases hydraulic conductivity and water storage capacity

Cover crops

- Reduce surface runoff and control erosion
- Add organic matter and improves soil structure and tilth
- Legumes fix atmosphere nitrogen
- Help to control weeds
- Increase soil productivity

The effect of Crop rotation

- Higher diversity in crop production
- Reduced risk of pest and weed infestations
- Greater disturbance of bio pores by diverse roots
- Better distribution and exploration of water and nutrients in soil profile
- Increased N fixation through legumes and improved balance of NPK from both organic and mineral sources.
- Increased humus formation

Controlling in field traffic

- The FAO now includes "controlling in-field traffic" as a component of conservation agriculture; this is accomplished by having field-traffic follow permanent tracks
- This can also be accomplished by using a ridge-till or permanent bed planting system rather than planting on the flat (Sayre & Hobbs, 2004)

Types of conservation tillage

1. Conservation tillage: Conservation tillage is a tillage system that creates a suitable soil environment for growing a crop and that conserves soil, water and energy resource mainly through the reduction in the intensity of tillage, and retention of plant residues.

Tillage practices such as ploughing, discing, harrowing and cultivation have an effect on soil aggregation, infiltration and resistance to wind movement. However, the soil erodability may occur due to destruction of soil structure by excessive tillage or improper conditions at the time of ploughing. Mulch tillage and contour tillage (contour farming) are the two types of tillage, which can conserve soil and moisture under dry land conditions.

a. Mulch tillage

Tillage that leaves the surface of the soil mulched with crop residues is called as mulch tillage. It is an effective measure to conserve soil and moisture since it,

1. Breaks and reduces the dispersing action of rain drops
2. Controls sheet erosion
3. Promotes infiltration of water

2. Minimum tillage: Aims at reducing tillage operations to the minimum necessity for a reasonably good seed bed to establish adequate crop stand and favourable soil conditions for growth and development of crops.

- Use of herbicide is must for control of weeds.
- Different types of minimum tillage are,
  - Row zone tillage
  - Plough pan tillage
  - Wheel zone tillage

3. Zero or No tillage: No-till systems, as the name implies, do not use tillage for establishing a seedbed. Crops are simply planted into the previous year's crop residue. No-till planters are equipped with coulters that slice the soil, allowing a double disc opener to place the seed at proper depth. The slot is closed with a spring press wheel. Herbicides are typically used as the sole means for weed control in no-till systems.

4. Mulch tillage: Mulch tillage system also called stubble mulch farming, is based on the principle of causing least soil disturbance and leaving the maximum of crop residue on the soil surface (30% more) and at the same time obtaining a quick seed bed. The soil is prepared in such a way that plant residues or other mulching material are specially left on or near the surface of farm which reduces the soil and water erosion.

5. Strip tillage system: The seedbed is divided into a seedling zone and a soil management zone. The seedling zone (5–10 cm wide) is mechanically tilled to optimize the soil and micro climate environment for germination and seedling establishment. The inter row zone is left undisturbed and protected by mulch. Strip tillage can also be achieved by chiseling in the row zone to assist water infiltration and root proliferation.

6. Ridge tillage: The soil is left undisturbed prior to planting but about one–third of the soil surface is tilled at planting with sweep or row cleaners; planting of row crops is done on formed cultivated ridges, while weeds are controlled by herbicides. Ridge tillage is a form of conservation tillage that uses specialized planters and cultivators to maintain permanent ridges on which row crops are grown. After harvest, the crop residue is left until planting time. The planter places the seed in the top of the ridge after pushing residue out of the way and slicing off the surface of the ridge-top.

The physical manipulation of surface soil should not destroy physical and chemical properties of soil. Tilth is the physical condition of soil in its relation to plant growth. In dry lands, tillage is required to:

(i) improve moisture conservation;
(ii) break hard pan and help root penetration;
(iii) control weeds;
(iv) provide better aeration and
(v) good seed bed. The soil aggregates stability depends upon the extent of disturbance by mechanical force during the course of tillage.

The concept of minimum and reduced tillage was more popular in western countries and highly mechanized plot. The concept of minimum or zero tillage may not work under Indian conditions (Table 93), except on Vertisols of certain areas and to certain extent under irrigation. Minimum tillage may not permit storage of as much rain water as possible in soil profile for subsequent use by the crop.

Table 93: Finger millet grain yield as influenced by tillage practices

Treatments	Grain yield (kg/ha)	Straw yield (kg/ha)	B:C Ratio
Conventional tillage	2432	5625	1.79
Reduced tillage	2102	5041	1.68
Minimum tillage	1987	4838	1.49
C.D. ($p = 0.05$)	229	270	

Source: (Anon., 2008)

Mechanization in Conservation Agriculture

- Reduced tillage and direct seeding with appropriate machines that reduce costs of labour, fuel and system of machinery.
- Increased opportunities for crop diversification, better response to market opportunities.
- Mechanical weeding and inter cultural operations in dry land crops reduce labour costs.
- Mechanical placement of fertilizers below the soil surface improves fertilizer use efficiency
- Reduced seed requirement
- Improved yields of all crops

Developing a crop residue management system to fit in one's farm operations requires careful consideration of factors such as,

- Capability of the planting machine to handle crop residue spreading of straw and chaff in the field
- Total crop residue and its condition, particularly after heavy crop or green and damp straw or lodged stalk
- Tall standing stubble reduces erosion but may cause plugging during planting
- A balance is needed between standing stubble and chopped residue on surface

The residues of crops act as a barrier for the flow of runoff and help in conserving soil moisture and provide more opportunity time for water to infiltrate into the soil. Adoption of this technology increased the yield of *rabi* sorghum and sunflower to an extent of 26 and 14 per cent, respectively over control in black soil (Table 94).

Table 94: Influence of crop residues on yield of crops

Crop	Yield (kg ha^{-1})		% increase yield over farmers practice
	Retention of crop residues	Farmers practice	
Rabi Sorghum	1606	1266	26
Sunflower	1299	1135	14

Source: Ramachandrappa *et al.* (2014)

Growing of horsegram residues as cover crop recorded significantly higher finger millet equivalent yield (2635 kg ha^{-1}) compared to field bean (2094 kg ha^{-1}) and control (1953 kg ha^{-1}), however horsegram as cover crop recorded significantly higher straw yield (3928 kg ha^{-1}), gross returns (₹66,507 ha^{-1}), net returns (₹37,607 ha^{-1}), benefit cost ratio (2.28) and rain water use efficiency (3.84 kg ha–mm^{-1}) compared to field bean and no cover crop (Table 95).

Table 95: Yield, economics and rain water use efficiency as influenced by conservation agriculture practices in finger millet+pigeonpea intercropping (8:2)

Treatments	Yield (kg ha^{-1})				Returns (₹ ha^{-1})		B:C ratio	RWUE (kg ha-mm^{-1})
	Finger millet grain	Straw	Pigeonpea grain	Finger millet equivalent	Gross	Net		
TILLAGE								
M_1	2338	4014	130	2565	65017	34430	2.12	3.74
M_2	2113	3492	117	2316	58498	28755	1.97	3.328
M_3	1644	2380	91	1802	45015	19039	1.73	2.63
S.Em.±	**42.95**	**210.45**	**6.75**	**51.28**	–	–	–	–
CD (p = 0.05)	**168.63**	**826.32**	**26.51**	**201.37**	–	–	–	–
COVERCROP								
C_1	1776	2827	102	1953	49160	20667	1.71	2.85
C_2	1910	3131	106	2094	52864	23951	1.83	3.05
C_3	2409	3928	130	2635	66507	37607	2.28	3.84
S.Em.±	**36.69**	**189.43**	**5.21**	**43.32**	–	–	–	–
CD (p=0.05)	**113.04**	**583.69**	**16.06**	**133.47**	–	–	–	–

Source: (Anon., 2018)

M_1: Conventional tillage (1 ploughing(tractor drawn) + 2 harrowing + 1 intercultural operations); M_2: Reduced tillage (1 harrowing + 1 intercultural operations + pre emergence herbicide); M_3: Zero tillage (pre emergence herbicide); C_1: Control (No cover crops); C_2: Field bean (HA–4) leaving the residue after harvesting the green pods for vegetables; C_3: Horse gram *in-situ* mulching.

Conservation Agriculture addresses agricultural production

CA has tremendous potential for achieving sustainable yield increases by improving the growth conditions for crops and the efficiency of input.

1. **Natural resource base**: CA reverses soil degradation processes and builds up soil fertility by facilitating better infiltration of rainwater and enabling the recharge of groundwater which reduces erosion and leaching and, in turn, water pollution.
2. **Biodiversity:** CA conserves and enhances biodiversity in the field.

3. **Labour shortage:** CA eliminates power-intensive soil tillage, thus reducing the drudgery and labour required for crop production by more than 50 per cent for small scale farmers. For mechanized farms, it reduces fuel requirements by 70 per cent and the need for machinery by 50 per cent.
4. Climate change livelihoods

Impact of Conservation Agriculture Practices

The rapid adoption and spread of CA technologies particularly zero-tillage for wheat is attributed to multiplicity of benefits. These include,

1. Reduction in cost of production
2. Reduced incidence of weeds
3. Saving in water and nutrients
4. Increased yields
5. Environmental benefits
6. Crop diversification opportunities
7. Enhancing productivity
8. Resource improvement

Benefits of Conservation Agriculture to Farmers

1. Reduced cultivation cost through savings in labour, time and farm power
2. Improved & stable yields with reduced use of input
3. In case of mechanized farmers, longer life and minimum repair of tractors and less water, power and fuel consumption
4. Benefits of CA come about over a period of time and in some cases, might appear less profitable in the initial years

Adaptation to climate vulnerability

- Planned adaptation is essential to increase the resilience of agricultural production to climate change
- Management practices that increase resilience and reduce yield variability under variable and extreme climate are,
  - Building resilience in soil
  - Adapted cultivars and cropping systems
  - Rainwater harvesting and recycling
  - Water saving technologies
  - Farm machinery (custom hiring) centers
  - Crop contingency plans

- Livestock and fishery interventions
- Weather based agro advisories
- Institutional interventions
- Village Climate Risk Management Committee (VCRMC)

The research, developmental and policy strategies suggested to restore and maintain soil quality on long–term basis by employing the principles of CA are,

- Checking soil resource through effective soil and water conservation (SWC) measures
- Rejuvenation and reorientation of soil testing program in the country
- Promotion of management practices which enhance soil organic matter
- Development and promotion of other bio-resources for enhancing microbial diversity and ensuring their availability
- Ensuring availability of balanced multi-nutrient fertilizers
- Enhancing the input use efficiency through precision farming
- Amelioration of problematic soils using suitable amendments and improving their quality to a desired level
- Land cover management
- Mass awareness about the importance of soil resource and its maintenance
- Need to constitute a high power body such as
  - National Authority on Land and Soil Resource Health
  - National Commission on Soil Resource Health

Steps to promote conservation farming

- Need to create awareness among the communities about the importance of soil resources, organic matter build up in soil.
- Systems approach is essential for fitting conservation tillage in modern agriculture.
- For the adoption of conservation tillage, complete package of practices may be identified for each agro ecological region.
- Use of herbicides has become inevitable for adopting, conservation farming – Scopes to study the allelopathic effects
- Low tillage, crop rotation, cover crops, maintenance of residues on the surface, control of weeds through herbicides must be studied in depth under diverse soil and climatic conditions across the country on long–term basis.
- Research focus is required on enhancing fertilizer use efficiency and reduction in use of pesticides.

- Appropriate time of sowing, suitable seed rate, depth of sowing, row orientation need research focus.
- Inter-disciplinary research efforts are required to develop appropriate implements for seeding in zero tillage, residue incorporation and inter-cultural operations.
- Research and management strategies to improve soil quality embedding conservation agriculture principles.

Challenges and Limitations of Conservation Agriculture

- Retention of residues for mulch cover
- Different spacing for rotation crops
- Unemployment to labour in first years
- Change of mindset

Practical benefits of Conservation Agriculture in Indo-Gangetic plains

The adoption of resource conservation technology has not only produced an additional 0.5 million tonnes of wheat, but also saved a foreign exchange of US$ 80 million through reduced fuel consumption in tillage and irrigation operations. Zero tillage has reduced the cost of cultivation by ₹2,500 ha^{-1} (US$ 55 ha^{-1}) and burning of fossil fuel by 50–60 L diesel ha^{-1}. Crop residues retained on the soil surface have improved the crop yield on an average by 247 kg ha^{-1} by reducing moisture loss and improving soil health.

Conditions to be met for adoption of Conservation Agriculture practices

- Rainwater harvesting/farm ponds/well recharging
- Initial investment has to come from the government by converging different schemes like MGNREGS, PKVY, NFSM, watershed programmes, *etc.*
- Operationalization of farm ponds needs to be done as a customized package for water harvesting and utilization (including inlet and outlet pitching and lining of the pond. water lifting, micro-irrigation system, *etc.*)
- Low-cost and easy to handle water-lifting devices and micro-irrigation system matching the needs of different categories of farmers need to be developed
- Farmers must be properly trained to handle and maintain micro-irrigation systems

Integrated nutrient management (conjunctive use of organic and inorganic fertilizers, residue management, mulching through biomass)

- Arrangements in place for capacity-building
- Policy favouring promotion of organic supply of nutrient in terms of crop residue management and higher biomass production for mulching, particularly in fragile soil environments

- Convergence with RKVY, NHM, *etc.,* may be useful
- Making biomass shredder available on subsidy to encourage mulching through agricultural residues (non-edible for animals). Rotavator may be used for incorporation of residue in the soil
- Location-specific on farm demonstration on a large scale
- Develop and provide power-operated machine to shred the non-edible (as fodder) agricultural residues like cotton stalk to use them for mulching

In-situ moisture conservation

- Identification of appropriate *in-situ* moisture conservation practices (ISMCP) on agro-climatic zone basis and further narrowing down to district sub-district level.
- Need to be implemented as an area approach by converging with relevant programmes like NREGA/RKVY/watershed programme covering all categories of farmers cultivating marginal or fragile lands or soils.
- Need to improve access to needed implements like ridge and furrow maker for wider scale adoption of ISMCPs through custom-hiring services promoted by self-help groups and Subsidy
- Large-scale on-farm demonstration
- Launching awareness campaign
- Appropriate implements (Bullock-drwan or tractor-drawn) need to be identified or modified or developed

Alternate land-use system (ALUS): Silvi-agri. Horti-agri and Agroforestry

- Capacity-building of the stakeholders
- Single-window delivery system of support for promoting ALUS starting from land preparation to the stage when ALUS begins to give economic benefits.
- Convergence among NREGA, NHM and watershed programme may help in this
- Delineate suitable areas for promoting ALUS considering local resources, traditional skills, market opportunities, fodder supply. Carbon credits and value-addition options

Chapter 11

Climate-Smart Agriculture

Climate change is looming large on the globe. Food security is one of the key issues that inevitably needs to be resolved under the specters of climate change, particularly in the fragile ecosystems. Never before in the history has Indian agriculture been as vulnerable and uncertainty ridden as it is today. A glimpse of the dynamics of Indian agriculture reveals that it has systematically deviated away from its base environment, the prop that nourishes all biological resources. Today's agriculture is valued against the prices it fetches from the market, especially the global market. Its contribution to human health and welfare, ecological integrity, resilience of nature *etc.*, are grossly neglected.

Agriculture in developing countries must undergo a significant transformation in order to meet the related challenges of achieving food security and responding to climate change. Projections based on population growth and food consumption patterns indicate that agricultural production will need to increase by at least 70 per cent to meet demands by 2050. Most estimates also indicate that climate change is likely to reduce agricultural productivity, production stability and incomes in some areas that already have high levels of food insecurity. Developing climate–smart agriculture is thus crucial to achieving future food security and climate change goals.

The production, processing and marketing of agricultural goods are central to food security and economic growth. Products derived from plants and animals include foods (such as cereals, vegetables, fruits, fish and meat), fibers (such as cotton, wool, hemp and silk), fuels (such as dung, charcoal and biofuels from crops and residues) and other raw materials (including medicines, building materials, resins, *etc.,*). Production has been achieved through a number of production systems which range from smallholder mixed cropping and livestock systems to intensive farming practices such as large monocultures and intensive livestock rearing. The sustainable intensification of production, especially in developing countries, can ensure food security and contribute to mitigating climate change by reducing deforestation and the encroachment of agriculture into natural ecosystems (Burney *et al.*, 2010 and Bellassen, 2010).

The climate changes impacts in agricultural sector needs to be addressed through mitigation and adaptive strategies (Fig. 26). The mitigation strategies include the selection of technologies that can help mitigate greenhouse gas

(GHG) emissions from the agriculture sector and promote a sustainable agriculture sector (Smith *et al.,* 2008). The adaptive strategies includes the location and crop specific contingencies. Various adaptation and mitigation strategies including use of climate resilient crops and cultivars for different regions are most essential for agriculture to successfully cope with climate variability. Improved agricultural practices evolved over time for diverse agro-ecological regions in India have potential to enhance climate change adaptation, if deployed carefully (Venkateswarlu *et al*., 2011).

A stronger focus on adapting agriculture to current and future climate change, for which simultaneous short term and long term strategies are needed. Short term responses include Real time contingencies to cope with seasonal weather aberrations for higher farm productivity, food security at house hold level and long–term strategies include mitigation of climate change besides having farm level impacts.

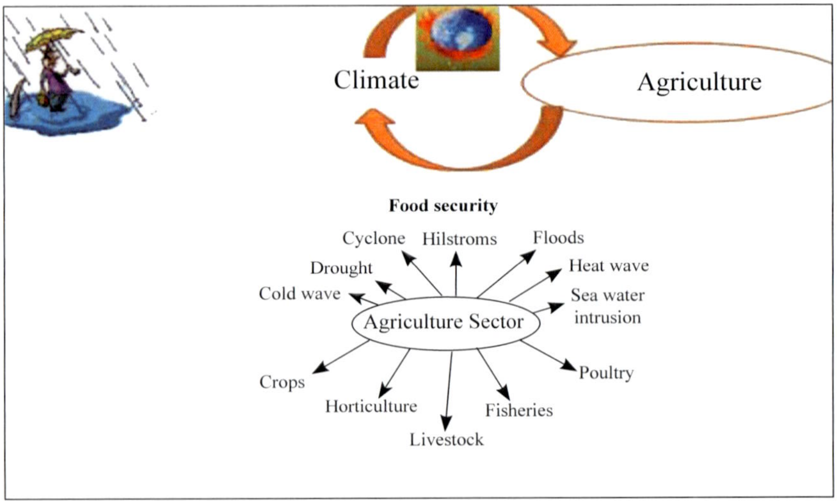

Fig. 23. Weather aberrations impact different sectors of agriculture

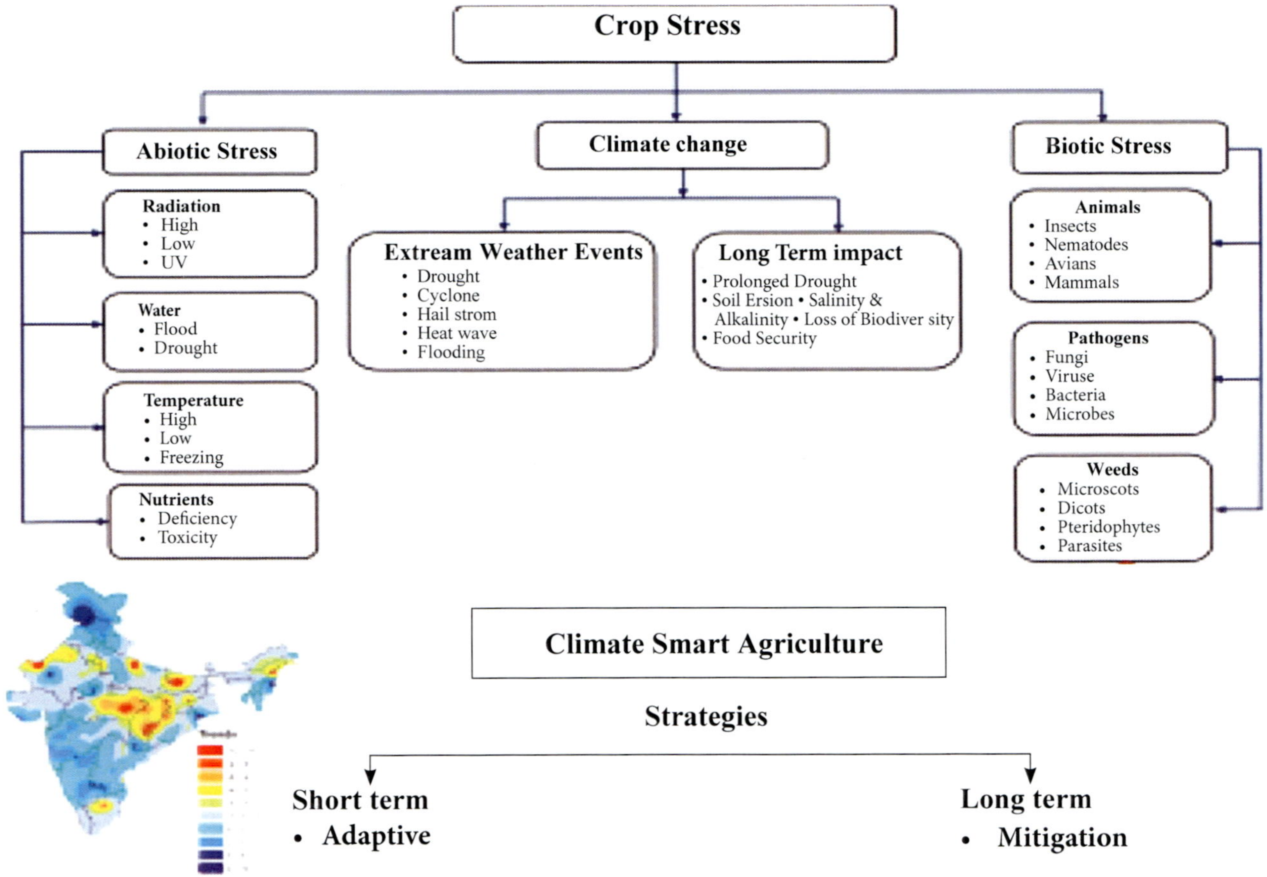

Fig. 24. Climate change impacts and strategies for climate smart agriculture

Long term strategies

1. Genetic resources: Genetic make-up determines plants and animals tolerance to shocks such as temperature extremes, drought, flooding and pests and diseases. It also regulates the length of growing season or production cycle and the response to inputs such as fertilizer, water and feed. The preservation of genetic resources of crops and breeds and their wild relatives is therefore fundamental in developing resilience to shocks, improving the efficient use of resources, shortening production cycles and generating higher yields (and quality and nutritional content) per unit area of land. Generating varieties and breeds which are tailored to ecosystems and the needs of farmers is crucial. The development and identification of climate resilient crop varieties, with enhanced tolerance to heat, drought, flooding, chilling and salinity stresses are essential in order to sustain and improve crop yields to cope with the challenges of climate change. It is essential to bridge the yield gaps, enhance the productivity and profitability, minimize risk and improve the livelihoods of millions of people dependent on agriculture. While, abiotic stresses such as drought, heat or cold may trigger a series of responses in plants that include changes in gene expression, signal transduction pathways, metabolic and molecular mechanisms as well as cumulative manifestations of these in terms of *Source* and sink relations for adaptation. Among various abiotic stresses, drought, heat, salinity, cold and flooding are the major factors that adversely affect plant growth and productivity (Maheswari *et al.,* 2012).

The need for stress tolerant cultivars has become paramount in the present context of climate change apart from various adaptation and mitigation strategies to feed the ever increasing population in the country. Concerted efforts of the National Agricultural Research System (NARS) during the last few decades resulted in development of stress tolerant varieties in several crops and efforts are further being strengthened to develop varieties tolerant to various abiotic stresses individually as well as those with multiple stress tolerance (Table 96). These stress tolerant cultivars can play an important role in coping with climate variability as well as enhancing the productivity (Maheswari *et al.,* 2015).

Table 96: Genetic resources-climate smart varieties

Crop	Variety	State
Rice	Sahabhagi	West Bengal, Bihar,
Wheat	PBW 396/527/644	Punjab, Haryana, UP, Rajasthan
Maize	Nityashree, Hema	Karnataka
Pearl millet	HHB 234/226/216	Rajasthan, Gujrath, Haryana
Barley	RD 2660	Rajasthan, UP, Haryana
Foxtail millet	RS 118, K–211–1	Karnataka

Crop	Variety	State
Finger millet	MR-1/6, ML 365	Karnataka
Sorghum	Pant Chari 5/7	Uttarakhand
Chickpea	RVG 202, JSC 55	MP, Maharashtra, Gujarat, UP
Field bean	HA 4	Karnataka
Horse gram	CRIDA 18R	AP, Karnataka, TN, Kerala
Pigeonpea	BRG–1/2	Karnataka, AP
Groundnut	GC 2/5	Gujrath

2. Agricultural biotechnology: Agricultural biotechnology stands out as a promising tool for the development of traits and varieties that help to mitigate and adapt to climate change. GM crops with pest resistance (Bt) and herbicide tolerance and conventionally bred varieties using marker selection in tissue culture have benefited agriculture by improving productivity and disease resistance. Biotechnological approach for reducing GHC through

- Efficient 'C' sequestering through higher biomass (Ainsworth *et al.*, 2008)
- Resistance to pest and diseases reduced the pest and pesticide load. Herbicide-resistance crops can be successfully cultivated under reduced tillage and save energy.
- Less photosynthetic migration to roots over shoots under irrigation.

3. Cover crops and Intercrops: These can protect soil from erosion, fix atmospheric nitrogen besides adding huge biomass. Cover cropping is an effective method of reducing emissions of CO_2. These crops grow over entire land areas or in localized spots such as grassed waterways, field margins, and shelterbelts. Compared to leaving fields fallow, they reduce emissions and can sequester carbon during periods when primary crops are not in the field (Lal, 1998). Cover crops are usually an option on surplus agricultural land or on cropland of marginal productivity.

Fig. 25. Covercrops and Intercrops

4. Nutrient Management: Efficient use of nitrogenous fertilizers can reduce N_2O emissions from agricultural fields. In addition, by reducing the quantity of synthetic fertilizers required, improved management can also reduce CO_2 emissions associated with their manufacture.

- Management of nitrogenous fertilizers: Fertilizer nitrogen management practices significantly influence the emissions of nitrous oxide in agriculture (Petersen *et al.*, 2006). These practices are fertilizer type, timing, placement, and rate of fertilizer application, as well as coordinating the time of application with irrigation and rainfall events.
- Use of Nitrification inhibitors: Emission of N_2O can be reduced by using nitrification inhibitors (Bhatia *et al.*, 2010), which slow the microbial processes that lead to N_2O formation. Ex: AM, N serve
- Slow release fertilizers: The use of slow release fertilizers offers a cost effective mitigation option. Slow release of urea and NH_4 based fertilizers can be achieved by using various coatings, chemical modifications, and changing the size of fertilizer granules. Ex: Urea super granules, Pellets (Travis Lybbert and Daniel Sumner, 2010).
- Synchronized nutrient application: Fertigation-Synchronous timing of N fertilizer application with N demand from plants is an important factor in determining the emissions of N_2O from row crop cultivation. Crop nitrogen intake capacity is generally low at the beginning of the growing season, increasing rapidly during vegetative growth and dropping sharply as the crop nears maturity.

5. Mitigation of C, N & P loss by mycorrhiza: One of the prime factors associated with enhancing soil carbon sequestration is the release of glomalin in mycorrhizal systems. Specific mycorrhizae: *Glomus intraradices, Glomus mosseae, Glomus fascicullatum, Glomus margarita* and *Glomus pellucida* have been reported to enhance soil carbon due to the release of glomalin. Glomalin is a glycoprotein that serves as gluing agent that facilitates soil aggregate formation and improves soil physical properties (Subramanian *et al.*, 2009). Enhancement of microbial biomass in the soil can also cause for improvement of soil physical and chemical properties.

6. Tillage: Conservation tillage is any method of soil cultivation that leaves the previous year's crop residue on fields before and after planting the next crop to reduce soil erosion and runoff, as well as other benefits such as carbon sequestration. Zero tillage and minimum tillage reduce the energy requirement and burning of petroleum products (Dinnes, 2004). Ridge till farming can also address the climate smart issues.

7. Bio-char: Biochar, a partially burnt organic matter used to improve agriculture and the environment in several ways and its stability in soil and superior nutrient retention properties make it an ideal soil amendment to increase crop yield. Biochar applications to soil sequester carbon and reduce emissions of non-CO_2 greenhouse gases. It also provides habitat for micro-organisms, which can increase soil microbial diversity. Biochar also acts as a soil conditioner that enhances plant growth, retains nutrients, and improves soil properties (Lehmann *et al.,* 2006). Bio-char is also referred as Black carbon.

8. Irrigation: All types of irrigation, such as flood, sprinkler, surface and sub-surface drip, can all enhance crop yields with subsequent increases in crop residues and enhanced carbon sequestration (Smith *et al.*, 2008).

Rain water harvesting *in-situ, ex-situ* through farm pond, bore well recharge and roof water harvesting, its efficient use through micro irrigation will enhance not only productivity but also addresses climate smart issues.

9. Manure and Bio-solid management

- Improved production, storage and Handling: Scientific production, covering and incorporation
- Off-field crop residue management: Vermi-composting
- Organic agriculture: Green manures & Bio-fertilizers
- Anaerobic decay of agricultural wastes: Household biogas digesters
- Bio-energy: Biofuels-trees, corn, sugarcane

10. Agro–Forestry including Horticulture: Agro-forestry is one of the important terrestrial carbon sequestration systems because of perennial nature. It involves a mixture of trees, agricultural crops, and pastures to exploit the ecological and economic interaction of an agro-ecosystem. The marginal lands have got very high potential for agro-forestry including horticulture. Agro-ecosystems play a central role in the global carbon cycle and contain approximately 12% of world terrestrial carbon (Youkhana and Idol, 2009).

11. Livestock management: Ammoniated straw and silage: The cellulose part of the straw can be digested and utilized by ruminant animals, while the lignin part cannot be digested. The main function of ammonization is to generate ammonolysis reaction using ammonia and straw, by damaging the ester bonds between lignin and polysaccharide, so that it can contact with digestive enzymes more easily, with an improvement in digestibility of straw. The digestibility and feed intake of ammoniated straw can be increased by approximately 20%, and the content of crude protein in ammoniated straw can be increased by two to three times (Guo, 1996).

Short term strategies

1. Fertilizer, manure, and straw management mitigation technology
   - Fertilizer types, amount, timing, proportion, slow releasing fertilizers, biofertilizers, green manures
2. Potassium fertilizer application technology (Babu *et al.,* 2006)
   - MOP application drop the redox potential, Inhibit methanogenic bacteria, Stimulate methanotropic bacterial population – Microbial biomass
3. Fertilizer use efficiency can be improved by fertilizer types, fertilizer nutrient ratios, fertilizer rates, timing of application, nitrification inhibitors, prilled urea + green manure, biofertilizers
4. Adoptive strategies: The important adoptive strategies that can address the climate smart issues *viz.* soil and water conservation, precision management practices, contingent crop and production practices, organic farming, biofertilizers and biopestcicides, need based mechanization for timely operation in dry land agriculture

a. Rainwater conservation and harvesting

Effective rainwater management is critical for drought mitigation and successful rainfed agriculture. *In-situ* soil and water conservation (SWC) practices improve soil structure and soil porosity, increase infiltration and hydraulic conductivity, and consequently increase soil water storage and promote crop growth and build adequate soil cover. Based on extensive research conducted at various locations in the country, several soil and water conservation measures have been recommended depending on the rainfall regimes (Table 97) and some of these practices are being adopted in a significant area in the country.

Table 97: Recommended soil and water conservation measures for various rainfall zones in India

Seasonal rainfall (mm)			
<500	**500–700**	**750–1000**	**>1000**
Contour cultivation with conservation furrows	Contour cultivation with conservation furrows	BBF (*Vertisols*)	BBF (*Vertisols*)
Ridging	Ridging	Sowing across slopes	Field bunds
Sowing across slopes	Sowing across slopes	Tillage	Vegetative bunds
Scoops	Tied ridging	Small basins	Level terrace
Tied ridges	Mulching	Field bunds	
Off-season tillage	Zingg terrace	Vegetative bunds	

Seasonal rainfall (mm)			
<500	**500–700**	**750–1000**	**>1000**
Inter row water harvesting system	Off-season tillage	Graded bunds	
Small basins	BBF	*Nadi*	
Contour Bunds	Inter-row water harvesting system	Zingg terrace	
Khadin	Modified contour bunds		
	Field bunds		

Source : Vittal *et al.* (2003)

In addition to *in-situ* conservation, efforts need to be made to divert the surplus water into storage structures which can be used either as stand alone resource or in conjunction with groundwater for meeting the critical irrigation requirements. In relatively high rainfall regions, the strategy is to conserve as much rainwater as possible and to harvest the surplus water for life saving irrigation and also for enhancing the cropping intensity and to maximize returns from the harvested water. Apart from enhancing the availability of water by various methods, increasing the water use efficiency should be the focus by arresting various kinds of losses associated with utilization of water.

b. Contingency crop planning

Contingency crop planning is essentially aimed at stabilization of crop output in the situation of late onset of monsoon, mid season and terminal drought. Contingency planning covers different drought situation *i.e.,* late onset of monsoon (delay by 2 weeks, delay by 4 weeks, delay by 6 weeks and delay by 8 weeks); mid season drought (at vegetative stage, at flowering or fruiting stage); terminal drought (early withdrawal of monsoon). The suggested drought contingency measures include change in crop or cropping system, crop management, soil nutrient and moisture conservation measures.

Any contingency measure, either technology related (land, soil, water, crop) or institutional and policy based, which is implemented based on real time weather pattern (including extreme events) in any crop growing season is considered as Real Time Contingency Planning (RTCP). If done timely and effectively, RTCP contributes to household and village food and fodder security (Table 98).

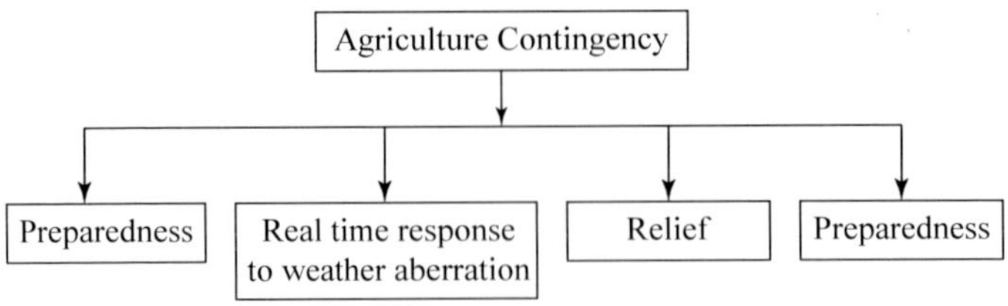

Fig. 26: RTCP implementation

Table 98: RTCP measures for various climate aberration in arable crops

Climatic aberration	RTCP measures
Delayed onset of monsoon	Beyond the sowing window, choice of alternate crops or cultivated crops or cultivars depends on the farming of situation, soil, rainfall and cropping pattern in the location and extent of delay in the onset of monsoon
Early season drought	• Re-sowing within a week to 10 days with subsequent rains for better plant stand when germination is less than 30%. • If the plant population is 50 to 75% of the optimum population (gaps in rows and also in between rows), sow the same crop of shorter duration variety in the gaps either within or in between rows. If less than 50% optimum population, sow suitable contingent crop as per the remaining effective growing season *e.g.,* pearl millet in place of sorghum up to 1"week of July • Thinning in small-seeded crops. • Interculture to break soil crust and remove weeds and create soil mulch for conserving soil moisture. • Avoid top dressing of fertilizers till favourable soil moisture. • Opening conservation furrows at 10 to 15 m intervals. • Ridge and furrow across the slope for effective moisture conservation as well in as rainwater in wide spaced crops (>30 cm). • Pot watering along with gap filling when the crop stand is less than 75% in crops like cotton. • Foliar spray of 2% urea during prolonged dry spells wherever ground/ surface water is available.
Mid-season drought	If symptoms such as drying of leaves, wilting of plants and cracking in black soils are observed, blade harrowing in rows during dry spell in coarse cereals, oilseed and pulse crops helps in creating dust mulch and closing of cracks in black soils. After relief of dry spell, open conservation furrows at 1.2 m distance, spray with 2% urea, particularly in pulse crops, castor and sunflower. Apply additional 10 kg N/ha. Plant protection Supplemental/ protective irrigation, if available. Repeated interculture to remove weeds and create soil mulch to conserve soil moisture. Avoid top-dressing of fertilizers until receipt of rains Opening conservation furrows for moisture conservation, Foliar spray of 2% KNO_3 or 2% urea solution or 1% water soluble fertilizers like 19–19–19, 20–20–20, 21–21–21 Opening of alternate furrows. Surface mulching with crop residues.

Climatic aberration	RTCP measures
Terminal drought	Providing life–saving or supplemental irrigation, if available. Harvesting crop at physiological maturity with some realizable yield or harvest for fodder. Prepare for winter (*rabi*) sowing in double-cropped areas Ratoon maize or pearl millet or adopt relay crops as chickpea, safflower, *rabi* sorghum and sunflower with minimum tillage after soybean in medium to deep black soils in Maharashtra Prefer contingency crops (horsegram/ cowpea) or dual-purpose forage crops on receipt of showers under receding soil moisture conditions.
Unseasonal	Re sowing Providing surface drainage. Application of hormones/nutrient sprays to heavy rainfall prevent flower drop or promote quick flowering/ fruiting and plant-protection measures against events pest/disease outbreaks with need based prophylactic/curative intervention At crop maturity stage, prevention of seed germination and harvesting of produce If Untimely rains occur at vegetative stage, the contingency measures include: Draining out the excess water as early as possible. Application of 20 kg N+10 kg K $acre^{-1}$ (0.4 ha) after draining excess water. Application of 50 kg urea 50 kg muriate of potash acre after draining excess water. Gap filling either with available nursery or by splitting the tillers from the surviving hills and Weed control. Suitable plant protection measures in anticipation of pest and disease outbreaks. Foliar spray with 1% KNO_3 or water-soluble fertilizers like 19–19–19, 20–20–20, 21–21–21 at 1% to support nutrition. Need-based fungicidal spray with Copper oxychloride 0.3% or Carbendazim 0.1% or Mancozeb 0.25% 2 to 3 times by rotating the chemicals Interculture at optimum soil-moisture condition to loosen and aerate the soil and to control weeds. Earthing up the crop for anchorage *etc.*
Floods	In sand-deposited crop fields/fallows, ameliorative measures include early removal or ploughing in of sand (depending on the extent of deposit) for facilitating *rabi* crop In silt deposited Indo-Gangetic Plains, early *rabi* crop plan is suggested in current cropped areas and current fallow lands. Draining out of stagnant water and strengthening of field bunds *etc.* In diara (flood prone) land areas, measures include alternate crop plan for receding situations Community nursery raising Re-transplanting in damaged fields and transplanting new areas or direct seeding depending on seed availability Prevention of pre-mature germination of submerged crop at maturity or of harvested produce by spray of salt solution.
Heat wave	Light frequent irrigation Foliar spray with thiourea or KNO_3 at recommended dose.
Cold wave	Light frequent irrigation during evening hours Basin mulching Smoking during night Application of supplemental fertilizers dose

Climatic aberration	RTCP measures
Frost	Light frequent irrigation during evening hours Basin mulching Smoking during night Application of supplemental fertilizers dose
Hailstorm	Use of anti-hail guns and anti-hail nets. Spray 0.1% Carbendazim to control secondary fungal infection. Open trenches to drain out excess water from the field, apply Dithane M-45 @ 0.2% on ear heads immediately after cessation of rains. If crop at harvesting stage, harvest mature cobs immediately and dry to avoid fungal growth. Store the cob and fodder after proper drying for 4–5 days, Foliar spray with sulphur @25 g per 10 litre water may be carried out to prevent its incidence.
Cyclones	Providing field drainage, staking and propping of plantation crops Cleaning and drying of harvested field crops

Source: Srinivasrao *et al.* (2016)

c Crop Management Practices

Table 99: Adoption of different crops and cropping systems in different soil types

Cropping systems	Shallow black soils and red soils	Medium deep black soils	Deep black soils
Sole Crops	Bajra, Redgram, Horsegram, sesamum, hibiscus, Castor and Sunflower	Ground nut, Redgram, Sunflower, Sesamum	Rabi Sorghum, Safflower, Chickpea and Sunflower
Intercroppings	Hy. Bajra+Red gram (2+1) Bajra + Castor (2:1) Groundnut + Redgram (4:2), Bajra + Groundnut (2:4)	Groundnut + Hy. Bajra (4:2) Ground nut + Red gram (4:2) Bajra + Castor (2:1) Bajra + Ground nut (2:4)	Safflower + Bengalgram (2:4) *Rabi* Sorghum + Bengalgram (2 :1)
Sequence cropping		Sunflower-Safflower Ground nut-*rabi* sorghum Green gram-*Rabi* Sorghum Green gram-Safflower Green gram- Sunflower	Sunflower-Safflower Greengram-Rabi Sorghum Greengram-Sunflower Green gram-Safflower
Crop rotation		Rabi Sorghum-Safflower + Chickpea (2 :4) Rabi Sorghum-Safflower Rabi Sorghum-Sunflower	

Source: Biradar *et al.*, 2016

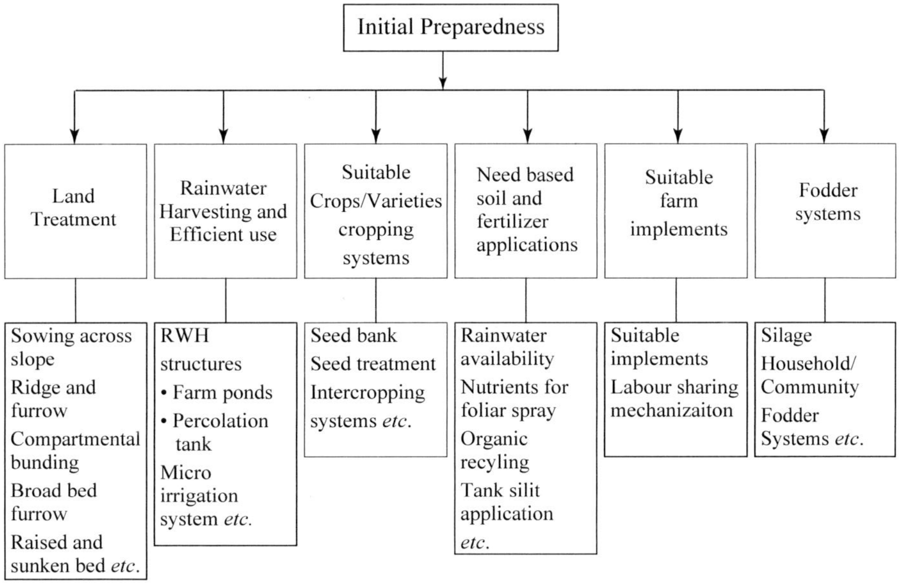

Fig. 29: Must-Do-Practices (MDPs)

- Land treatment practices include *in-situ* soil and water conservation measures *viz.,* Sowing across slope, Ridge and furrow, Compartmental bunding, Broad bed and furrow, Raised and sunken bed *etc.*
- *Ex–situ* rainwater management and efficient use measures *viz.,* Farm ponds, Percolation tank, Micro irrigation systems *etc.*
- Precision management practices *viz.,* seed treatment, optimum spacing or plant population, timely nutrition and plant protection measures *etc.*

d. Soil management

Soils hold the key to productivity and resilience to climate vagaries in rainfed agriculture in India. Loss of fertile soil as erosion, depletion of soil organic matter, emerging secondary and micronutrient deficiencies, soil compaction, surface crusting, loss of soil biodiversity have become potential limiting factors for productivity enhancement in these regions. Hence, these factors need improvement in facing the climate adversities. Based on 16 long-term manurial experiments under rainfed conditions in AICRPDA network, it was showed that each ton of soil organic carbon improved productivity of rainfed crops by up to 0.15 t ha^{-1} $year^{-1}$. Location specific integrated nutrient management (INM) practices were identified and being promoted based on locally available organic resources. Balanced nutrition particularly optimum potassium nutrition also contributes to mitigation of water stress conditions as K controls water relations in plant growth. Foliar sprays of KNO_3, and thiourea during mid-season droughts reduce the negative effects of droughts in several rainfed crops (Gopinath *et al.*, 2016).

Balanced Nutrition helps in not only higher productivity of crops but also saving of fertilizers and increase in the benefit cost ratio. It is a climate smart technology giving bost to crop growth and sustainance of crop during drought due to balanced fertilizer application. Experiences of National Innovations on Climate Resilent Agriculture Project of CRIDA at Nelamangala, Bangalore rural District, Karnataka revealed increased productivity, net returns and Benefit: Cost ratio of Ground nut + Pigeon pea and Finger millet + Pigeon pea due to application of fertilizers based on soil test results compared to blanket recommendation (Table 100).

Table 100: Balanced fertilizer recommendation for improving soil fertility and enhancing crop productivity

Cropping system	Blanket Rec. (NPK kg ha^{-1})	Based on soil test (NPK kg ha^{-1})	Yield (kg ha^{-1})	Net returns (₹ ha^{-1})	BC ratio
GN + PP	—	37.5:50:12.5	1333 (33%)	48,700	2.71
	25:50:25	—	1000	32,500	1.86
FM + PP	—	62.50:27.50:50	1995 (87%)	22, 579	2.06
	50:40:37.5	—	1065	4,095	1.22

Source: (Ramachandrappa *et al.*, 2014)

- Contingent crops and production practices emphasizing suitable crops or varieties, cropping systems through seed bank, intercropping systems *etc.*

e. Delayed onset of Monsoon-Change in Varieties

Table 101: Crops and Varieties for weather aberrations at different locations

Centre	Delay in onset by	Crop	Variety	Yield (kg ha^{-1})	Net returns (₹ ha^{-1})	RWUE (kg ha-mm^{-1})
Chianki (Loamy) SRF: N- 1038 mm; A- 904 mm)	21 days (28 June)	Rice Sesame	IR 92521-7-5 Shekhar	2220 644	21520 19268	2.55 2.75
Rewa (Deep black) SRF: N- 965 mm; A- 858 mm)	10 days (2 July)	Soybean Pigeonpea	JS 20–29 Asha	1016 1780	29720 64100	2.85 5.00
SK Nagar (Sandy/ sandy loam SRF: N- 598 mm; A- 931 mm)	31 days (26 July)	Pearlmillet Clusterbean	GHB 558 GG2	445 325	2070 7520	0.25 1.43

SRF– Seasonal rainfall; N - Normal; A- Actual

Contingent crops for delayed sowing

Table 102: Small millets as contingency crops

Year	Yield (kg ha^{-1})			
	Foxtail millet	Little millet	Kodo millet	Proso millet
2011	729	986	**2508**	494
2012	486	**972**	368	226
2013	1579	1080	869	659
2014	970	739	**1247**	923
2015	933	923	1173	914
Average	**939**	**940**	**1233**	**643**

Short duration pulses as contingency crops

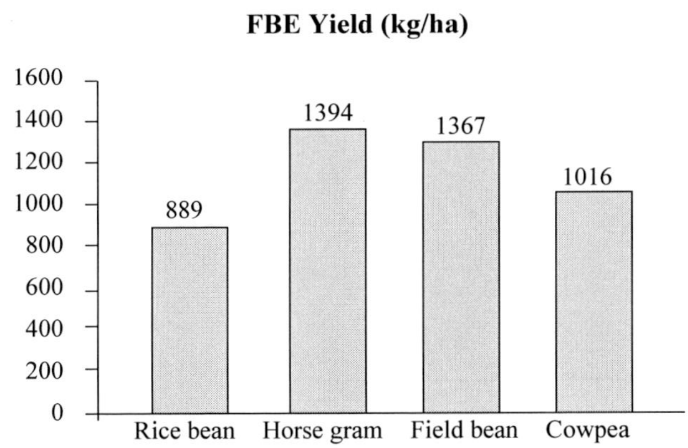

FBE – Field bean equivalent yield

f. Fodder security for animals under IFS approach through silage, hydroponics *etc.*, IFS plays an imperial role for maximizing farmers profit and production to meet the nutritional requirement with food security with less investment. Further, IFS is more advantageous for those farmers, who are able to produce more by using optimal resource utilization and recycling of waste materials and family labour employment. The regular flow of income from other subsidiary enterprises like dairy, sheep, goat, piggery, apiculture, mushroom cultivation, sericulture in IFS approach will help the farmer to overcome crop failures due to vagaries of monsoon and can sustain his livelihood. The components of IFS (Crop, Animal, Sericulture, Fishery, Fodder, Mushroom, Bio-gas *etc.*) will be decided based on the available bio-physical components of the farmers in the locality.

g. Emphasis for organic farming, bio-fertilizers, bio-pesticides (bio control agents)

h. Mechanization for timely operation (Custom hiring)

The agriculture sector is a significant contributor to GHG emissions and requires major consideration in global mitigation efforts. A lack of awareness and guidance as well as lack of economic strength of farmers have led to continuation of older and higher GHG-emitting agricultural practices. Land management practices, bio-energy plantation and utilisation, reduced tillage farming and other soil organic carbon management should be adopted to promote mitigation besides, appropriate location specific adoptive measures. Climate change is inevitable and likely to impact but, need is for the seed of mindset to manage key agricultural activities to make climate smart farm and nation.

(i) Method of establishment in finger millet

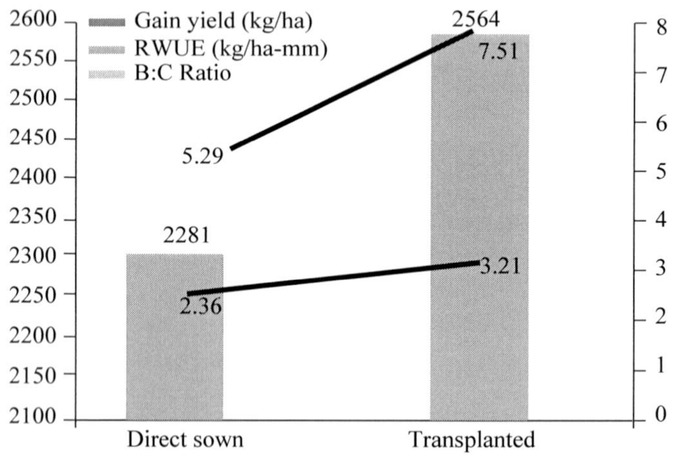

Transplanting under delayed rainfall is a realtime contingency measure which gives higher productivity and economic returns compared to direct seeding.

j. Real time agro-advisory services

- Agromet-advisories and crop-weather bulletins-twice a week (Tuesday and Friday) in collaboration with AICRPAM and IMD.

Agro-advisories

- Messages written in front of milk collection
- During the year 2015, there was forecast on cyclonic rainfall in the month of November and crop weather advisory suggested farmers not to harvest the crop in the first and second week of November.
- Few farmers harvested finger millet and undergone crop losses.

Standing crop

Harvested crop germinated due to continuous rainfall

Contingent Crop Plan for aberrant weather situations

Selection of crops depends on under contingent crop planning depends on

i. Commencement of rains may be quite early or considerably delayed
ii. There may be prolonged breaks during cropping season
iii. There may be spatial and temporal aberrations
iv. Rains may terminate considerably early or continue for longer periods

When the rains are delayed

- Use short duration varieties
- Sow sunflower, cowpea or horsegram instead of ragi
- Transplant seedlings instead of direct seeding
- Dry seeding when the land is ready but dry at normal sowing seasons
- Increase the seed rate to get more population as late sown ones do not makes much growth

If rain stops after germination

- Intercultivate to remove weeds
- Remove week seedlings and reduce population

- Irrigate from farm pond water
- Gaps fill after rains revive and top dress with nitrogenous fertilizers

If rain stops before stipulated time

- Irrigate from farm ponds
- Intercultivation to reduce evaporation
- Mulching with organic residues, including removed weeds from the fields

Table 103: Climatic extremes and measures

S. No.	Month	Climatic extremes	Measures
On farm			
1.	June	Hailstorms affected the germinated pigeon pea	Gap filling by resowing
2.	June 2nd fortnight	Dry spell	1. Favourable for groundnut for flower bursting
			2. Opening of conservation furrow between two rows of Pigeon pea in Groundnut + Pigeon pea (8:2)
3.	July	Delay in the monsoon	Five farmers have taken up finger millet nursery (MR-1) and transplanted after 21days after the receipt of rains.
4.	July 2nd fortnight	Delay in the monsoon	Medium duration varieties of finger millet like GPU–28 were given to 13 farmers.
5.	August1st fortnight	Delay in the monsoon	Short duration varieties of finger millet like GPU–48 was given to 3 farmers
		Dry spell	Thinning, gap filling and weeding was done in the early sown finger millet.
6.	August2nd fortnight	Excess run off rain water	Collected in the farm pond and used to nourish vegetable plants, fruit plants *etc.*
7.	Oct-Nov	Rains	Harvesting ear heads of finger millet

Chapter 12

Watershed Management

Watershed is defined as a natural hydrological unit that covers a specific land surface area from which runoff passes through a common outlet or area with common drainage or outlet. In simple terms, it implies a catchments or drainage basin from which water drains towards a single channel. It may extend over a few acres only or may cover thousands of acres.

Watershed management is the integration of technology within the natural boundaries of a drainage area for optimum development of land, water and plant resources to meet the basic minimum needs of the people in a sustained manner.

It is also defined as the development and management of watershed resources in such a manner as to achieve optimum production without deterioration of resources base or disturbing the ecological balance. It is also termed as "Resource centered technology" since it helps in assessment, augmentation and optimal utilization of all natural resources *viz*., land, water and vegetation, it prevents deterioration of resources and at the same time ensures sustained productivity of land to meet basic needs of people.

Need and advantages

1. It is an acceptable basic hydrological unit of planning for optimum use and conservation of soil and water resources.
2. Here, development is not confined to agricultural land alone but covers the whole land area starting from the highest point of the watershed (ridge line) to the lower most point of outlet at the bottom of the slope. It means every part of land including barren, sloppy and marginal lands being treated as per its capability *i.e.,* integrated treatment of arable & non arable lands.
3. It aims at comprehensive development of all resources in the watershed *i.e*., holistic. It starts from the most important resources *viz.,* soil and water and extends to other resources like crops, trees, livestock *etc.,* Hence, Watershed approach is considered as Holistic approach.
4. It is a multidisciplinary approach involving scientists from all related disciplines of Agronomy, Engineering, Horticulture, Forestry, Soil Science, Extension, Economics, *etc.*

5. It provides for involvement of farmers in planning, execution and monitoring of the development.

Principles of watershed management

The main principles of watershed management based on conservation, generation and utilization of resources are:

1. Utilizing the land according to its capability
2. Protecting productive top soil
3. Reducing siltation hazards in storage tanks and reservoirs and lower fertile lands
4. Maintaining adequate vegetation cover on soil surface throughout the year
5. *in-situ* conservation of rain water
6. Safe diversion of excess water to storage points through vegetative waterways
7. Stabilization of gullies by providing checks at specified intervals and thereby increasing ground water recharge
8. Increasing cropping intensity through appropriate cropping
9. Safe utilization of marginal lands through alternate land use systems with agriculture - horticulture - forestry - pasture systems with varied options and combinations
10. Water harvesting for supplemental and off–season irrigation
11. Maximizing agricultural productivity per unit area per unit time and per unit of water
12. Ensuring sustainability of the ecosystem befitting the man-animal-plant-water complex
13. Maximizing total income and improving the socioeconomic status of the farmers

Objectives of watershed management

The main objectives of this multipurpose programme are as follows

a. To conserve the basic resources of soil, rain water and vegetation
b. To achieve higher biomass production both in arable and non-arable areas.
c. To impart stability to crop yields through proper rainwater management, establishing good cropping patterns and ensuring proper land use.
d. To enhance the income of individuals through adoption of alternative enterprises.
e. To restore ecological balance.

Components of watershed approach

1. Water resource improvement
2. Soil and moisture conservation in cultivate lands
3. Land treatment in non-arable lands
4. Improved cropping
5. Alternate land use system and integration of livestock in farming system

Aims of watershed approach

1. Increased land productivity through improved technology
2. Sustaining the resource base thorough improved conservation measures
3. Augmentation of resource base *viz.,* soil productivity and water availability

Action plan/steps in watershed development

1. **Identification and selection of watershed**: The boundary of the watershed has to be marked by field survey starting from the lowest point of the watercourse and proceeding upwards to the ridge line. The area may vary as low as 100 ha to as high as 10000 ha.
2. **Description of watershed:** Basic information like Location, Area, shape and slope, Climate, Soil - geology, hydrology, physical, chemical and biological properties, erosion level, Vegetation, Land capability, Present land use pattern, Cropping/Farming system adopted, Man power resource, Socio-economic data, Infrastructural and institutional facilities *etc.,* has to be collected.
3. Analysis of problems and identification of available solutions
4. Designing the technology components
   a. Soil and moisture conservation measures
   b. Run off collection, storage and recycling
   c. Optimal land use and cropping system
   d. Alternate land use system and farming system
   e. Other land treatment measures
   f. Development of livestock and other allied activities
   g. Ground water recharge and augmentation
5. Preparation of base maps of watershed incorporating all features of geology, hydrology, physiography, soil and proposed development measures for each part of watershed.
6. Cost-benefit analysis to indicate estimated cost of each component activity, total cost of project and expected benefit.
7. On-farm research to identify solutions for site-specific problems.

Watershed Characteristics

The following watershed characteristics, which effects runoff and soil loss

I: Climatic characteristics

If the climatic condition is dry before the rainfall, infiltration and evapotranspiration of rainwater is more. The more the intensity of rainfall in the watershed, the more is the peak flow disposal from the area. If the duration is less, time taken to dispose with lower peak flow is also less.

II. Physiographic Characteristics

1. **Shape:** Watersheds differ in their shape based on morphometric parameters like geology and structure. The shape of watershed has a dominant effect on the characteristics of the hydrograph of the watershed such as peak flow, overland flow, and base of hydrograph (run-off time).
2. **Size:** The runoff rate and volume increases with increase in watershed area. Watersheds are also classified into different categories based on area that the watershed contains is as follows:

Size (ha)	Classification
50,000–2,00,000	Macro-watershed
10,000–50,000	Sub-watershed
1,000–10,000	Milli-watershed
100–1,000	Micro-watershed
10–100	Mini-watershed

Source: Singh,1994

3. **Slope:** It controls the rainfall distribution and movement, land utilization and watershed behavior. If slope is more, the velocity is more, and hence flood water drains quickly results in higher peak discharge reduces infiltration increases the soil transportation.
4. **Drainage:** The stream order, drainage pattern, and drainage density have a profound influence on runoff, infiltration, land management *etc.* It determines the flow characteristics and erosional behavior. If drainage density is more, peak runoff is more.
5. **Land Use:** The land in watershed is used for numerous purposes such as - dwelling houses, cultivation, livestock, water harvesting, roads and railways. Land use affects the run-off rates.
6. **Vegetation Cover:** Vegetation cover retard the flow and increases the infiltration and interception, there by reducing the peak flow. Vegetation cover protects the soil erosion. Detailed information on vegetation helps in choosing type, mode and manner of greening the watershed.

7. **Soil:** The soil such as sand absorbs a larger part of run-off water, so the rate of flow is less. Clay soil produces more flow as its absorption is less. Soil parameters such as depth, nature, moisture and fertility determines crops.
8. **Geology:** Rocks and their structure control formation of a watershed itself because their nature determines size, shape, physiography, drainage, and groundwater conditions. Geological conditions affects the run-off of watershed. If there are some cracks in rocks joining surface water and groundwater, a part of water flows quickly underground. Lakes, Storages, Swamp hold a part of excess rainfall.
9. **Hydrology**: The availability, quality and distribution of surface water is basic to the final goal of growing vegetation in a watershed. Hydrological parameters such as intensity and duration of rainfall help in quantification of water available in watershed. If rainfall continues for longer period with low intensity, evaporation and infiltration increases but run-off is less. Further it may cause landslide in sloping hillside areas.

III. Socio-economic characteristics

1. **Demographic profile:** Total population of the area, sex, age, ethnic group stratification, occupation and migration will decide the characteristics of watershed.
2. **Sociological stratification:** Caste, families, authorities, political organization, social status and political role of individual members and spatial distribution of the community will decide the watershed characteristics.
3. **Farm structure:** Area of farm, number of farm workers, type of farming and crop production technologies followed will decide the characteristics of watershed.
4. **Land tenure systems:** Size of land holding, landholding class and castes, methods of land acquisition, alienation of land rights, relationship between land rights and social, political or religious systems, land leasing, sharecropping and marketing system of agricultural produce affect watershed functioning.

Causes for Watershed Deterioration

Watershed degradation is the loss of value over time, including the productive potential of land and water, accompanied by marked changes in the hydrological behaviour of a river system resulting in inferior quality, quantity and timing of water-flow. Watershed degradation results from the interaction of physiographic features, climate and poor land use (indiscriminate deforestation, inappropriate cultivation, disturbance of soils and slopes by mining, the movement of animals, road construction, and badly controlled diversion, storage, transportation and use of water). Watershed degradation, in turn, leads to accelerated ecological degeneration, reduced economic opportunities and increased social problems. The activities responsible for watershed deterioration are as follows:

1. **Cultivation of crops:** Cultivation of crops without proper soil and water conservation *viz.*, cultivation on slopy land, cultivation along the slope, cultivation of erosion permitting crops and leaving land fallow during off season.
2. **Over grazing of grasslands:** Uncontrolled grazing and development of cattle tracks leads to poor vegetation and compaction of soil, resulting in lower infiltration rates and high runoff.
3. **Deforestation:** Excessive cutting of timber yielding trees and cutting of forest trees for firewood leads to less infiltration and more runoff. Uncontrolled grazing which inhibits regeneration from seed and coppicing, cutting and felling of trees on steep slopes leads to soil erosion and silting up of the tanks.
4. **Shifting cultivation:** Jhum cultivation has now become the major cause of forest loss in hilly watershed. This is because of the fact that additional profits are made from firewood and charcoal production in the field thereby totally wiping out for easy regeneration of trees growth. The increasing population with the unceasing jhumming of the land by every household every year to feed their families is the main economic factor behind deforestation.
5. **Mining and construction activities:** Mining and construction activities destroy the vegetation and the landscape, which leads to low infiltration and more runoff.
6. **Community participation:** The communities in the watershed are not actively participating to achieve the main principles of watershed, which leads to soil erosion.

Watershed Planning

A watershed plan is a strategy and a work plan for achieving water resource goals that provides assessment and management information for a geographically defined watershed. It includes the analyses, actions, participants and resources related to development and implementation of the plan. The watershed planning process uses a series of cooperative, iterative steps to characterize existing conditions, identify and prioritize problems, define management objectives, and develop and implement protection or remediation strategies as necessary.

The main components in a watershed planning includes *viz.*, public concerns, watershed inventory (water quality, physical, and social data), problem identification, identify *Source*s of problems, selection of critical areas, set goals, and objectives and measure success. The primary purpose of a watershed management plan is to guide watershed coordinators, resource managers, policy makers, and community organizations to restore and protect the quality of lakes, rivers, streams, and wetlands in a given watershed. The plan is intended to be a practical tool with specific recommendations on practices to improve and sustain water quality. These are also "living documents", meaning that as conditions change over time in a watershed, the plan must be reexamined and revised to reflect goals that have been achieved or not met.

Watershed Management in Drylands

Dry regions cover about one-third of the earth's land surface, and slightly over half of this area is inhabited by more than 850 million people. The remainder is climatically so arid and unproductive that it cannot adequately support human life. But the degradation of the land and water resources of arid zone watersheds by human activities is turning potentially productive dry lands into unproductive deserts in Asia, Africa and Latin America. This process is called desertification. It has been estimated that a collective area larger than Brazil, with rainfall above the level classified as semi-arid, has been degraded to desert-like conditions. This does not take into account the far greater degradation that is taking place within the potentially productive semi-arid zones.

Table 104: Finger millet yield and economics under different crop production technology

Practice	Local variety			HYV		
	Yield ($q\ ha^{-1}$)	**Net returns (₹ ha^{-1})**	**B:C ratio**	**Yield ($q\ ha^{-1}$)**	**Net returns (₹ ha^{-1})**	**B:C ratio**
Farmer's practice	4.12	–592.6	0.67	5.35	–363	0.8
Contour cultivation	4.71	–462.8	0.71	6.4	–90	0.95
Use of 50% rec. fertilizers	5.05	–458.0	0.77	8.75	372	1.18
Contour cultivation + 50% rec. fertilizers	6.85	–67	0.97	10.6	834	1.3
Contour cultivation + RDF	8.08	–61.4	1.03	12.95	1281	1.54

Source: Krishnappa *et al.*, 1994

Present land use patterns on arid zone watersheds must be reshaped in order that delicate water relations are not pushed beyond their limits. As the number of people and animals living in the arid zones climbs and the quality of the land on which they must live simultaneously declines, the impact will be global unless solutions are implemented. The different agronomic practices are required to conserve the soil and moisture and in turn sustain the productivity in drylands.

In the watershed area contour cultivation with 100% RDF recorded 8.08 and 12.95 q ha^{-1} of fingermillet grain yield in local and HYV, respectively with B:C ratio of 1.03 and 1.54, respectively (Table 104). Similarly, in case of groundnut contour cultivation with RDF registered 11.25 q ha^{-1} of pod yield and 2.02 B:C ratio (Table 105).

Table 105: Groundnut yield and economics under different crop production technology

Practice	Yield (q ha^{-1})	Net returns (₹. ha^{-1})	B:C ratio
Farmer's practice	5.3	607	1.23
Contour cultivation	6.96	1522	1.27
Use of 50% rec. fertilizers	7.3	1421	1.48
Contour cultivation + 50% rec. fertilizers	8.2	1916	1.63
Contour cultivation + RDF	11.25	3428	2.02

Source: Krishnappa, *et al.*, 1994

Finger millet + Pigeonpea (10:2) cropping system with moisture conservation furrow between 2 rows of pigeonpea recorded higher returns of ₹14, 918 per hectare with B:C ratio 2.26 as compared to farmers' practice (1.15) despite, long dry spells of 25 days during grain filling stage at Nelamangala under Operational Research project, Bangalore (Ramachandrappa *et al.*, 2010).

Watershed Management Through Remote Sensing and GIS use of Thematic Maps

The growing pressure on land for food, fuel, fiber and fodder due to burgeoning population have resulted in competing and conflicting demands on finite natural resources. An estimated 175 million hectares of land in India, constituting about 53 per cent of total geographical area suffers from the land degradation process. For a country which supports 15 per cent of world population on 2 per cent of global land area, the problem is serious enough. The water and land resources need to be developed, utilized and managed in a holistic approach.

In Karnataka, dryland farming is practiced in about 78 per cent of the net cultivated area, crop production in these areas is critically dependent on uncertain and erratic rainfall. With land degradation, low soil fertility, lack of infrastructure facilities relating to processing and marketing in rural areas, improper management of common lands, frequent occurrence of drought, shortage of drinking water and assured irrigation to crops, the productivity levels are gradually declining.

In order to overcome these constraints and problems, integrated watershed development approach needs to be adopted. By following appropriate soil and water conservation measures in the catchment with optimal harvesting of runoff, initiation of recharging mechanisms with proper exit protocol, further spread of land degradation can be minimized which in turn improves soil fertility and drinking water problem to a considerable extent. Through proper awareness programme and inter departmental linkages, marketing mechanism can be provided.

In reality however, gross neglect of recharge and over exploitation of under groundwater resource has generally led to continuous dipping in the dynamic water table in most of the places. While increased protective irrigation of rainfed land is absolutely essential for achieving higher yields, it is important to recognize that groundwater irrigation is roughly twice as efficient as surface water or canal irrigation. The major issues to be tackled in conserving and efficient management of these natural resources are erosion, drainage, irrigation, moisture conservation, impounding rain water and overall water resource development of the area.

Hence, in an integrated approach, the land use, soil, slope, climate and relevant parameters are considered for suggesting the integrated resource development plan showing soil and water conservation measures and perennial crop cover in planning appropriate agro forestry, agro horticulture and pastural systems.

History and development of remote sensing and GIS in India

The word 'Remote Sensing' was coined by Evelyn J Fruit and it was first used in the United States in the 1960's and encompassed photogrammetry, photo-interpretation, photo geology *etc*. Remote sensing is the science and art of obtaining information about an object, area or phenomenon through the analysis of data acquired by a device that is not in contact with the object, area or phenomenon under investigation.

Remote sensing was one of the space applications to be put to use in India. An aerial survey using infra-red film was carried out from a helicopter in 1970 to study root-wilt disease of coconut plantations in Kerala. In 1962, Telstar, the world's first communications satellite was launched by the US. Dr. Vikram Sarabhai immediately recognized the potential of a communications satellite to reach out to not only far corners of a country but across countries and continents. In 1967, the space department conducted a joint study with NASA to examine alternatives for television coverage in India. It concluded that a hybrid system involving direct broadcast through a satellite and retransmission through conventional terrestrial TV stations would be the most effective path.

Later Indian Space Research Organisation (ISRO) established the LANDSAT (Satellite Launched by America) receiving station (receives information from the satellite) at National Remote Sensing Agency (NRSA), Hyderabad in 1979 which provided valuable experience in the interpretation and analysis of space based data.

Following the successful demonstration flights of Bhaskara 1 and Bhaskara 2 launched in 1979 and 1981, respectively, India began development of an indigenous IRS (Indian Remote Sensing) program to support the national economy in the areas of "agriculture, water resources, forestry and ecology, geology, water sheds, marine fisheries and costal management". Remote sensing applications in the country, under the umbrella of NNRMS, now cover diverse fields such as crop acreage and yield estimation, drought warning and assessment, flood control and damage assessment, land use/land cover information, agro-climatic planning, wasteland management, water resources management, under-ground water exploration, prediction of snow–melt run off, management of watersheds and command areas, fisheries development, under development, mineral prospecting forest resources survey, active involvement of the user ministries/departments has ensured in an effective harnessing of the potential of space–based remote sensing. An important application of IRS data is in Integrated Mission for Sustainable Development (MSD) initiated in 1992. IMSD, under which 174 districts have been identified, aims at generating locale–specific action plans for sustainable development.

The first two IRS-1A (March, 1998) and IRS-1B (August, 1991) were launched by Russian Vostok boosters from the Baikonur Cosmodrome. IRS-1A failed in 1992, while IRS-1B continued to operate through 1999. Then, IRS-1A and IRS-1B were to be joined in 1993 with IRS-1E, when its PSLV Launch vehicle failed to reach Earth orbit. Thirteen months later, in October, 1994, the PSLV functioned correctly, allowing IRS-P2 to assume an 820-km, sun-synchronous orbit. IRS-P2 carried the LISS-2 system. IRS-1C, successfully launched on December 28, 1995 on board a Molniya rocket of Russia, IRS-P3 was launched by PSLV in 1996.

The Indian Space Research Organisation (ISRO) and its commercial marketing arm, ANTRIX crop. Ltd., successfully launched the IRS-ID earth imaging satellite on 29 September 1997 from Sriharikota, india. The satellite is an identical twin to the IRS-1C. The dual use of these satellite provide 5.8 meter resolution images to customers twice as often as was possible with just the IRS-1C, IRS P4 in 1999, Resourcesat 1 in 2003, Cartosat 1 in 2005, Cartosat 2 in 2007, Resourcesat 2 in 2011 and Cartosat 3 in 2019.

Geographic information system which consists of three words, *viz.* Geographic, Information and System. GIS can be defined as a set of tools for collecting, storing, retrieving, transforming and displaying geographically referenced spatial data with its corresponding attribute information to meet a specific requirement.

Geographical information system has been developed during mid 20th century with the development of computer science. The first true operational GIS was developed by Dr. Roger Tomilson, Department of Forestry and Rural Development, Canada. It was called as Canada Geographic Information system (**CGIS**) and who is also known as the 'Father of GIS'.

Application of remote sensing and GIS in watershed management

Watershed prioritization and characterization

Watershed prioritization

It is always better to start management from the highest priority sub watershed. That is from technical sustainability point of view treatment of watershed should start from ridge to valley. Watershed prioritization is, thus the ranking of different sub watershed according to the order in which they have to be taken up for treatment. High priority is one which has abundant severity problems like erosion, gravelliness, sub surface drainage *etc*., need immediate treatment for arresting further spread of gullies and to control the accelerated surface soil erosion. Medium priority areas are those with moderate problems, where certain measures are implemented and low priority is one where minor problem of soil erosion, soil depth, soil texture are noticed and can be gradually treated depending on the resources.

Watershed prioritization is made using satellite remote sensing data as the main input. Available summer season data has to be used to know the area under waste lands, degraded lands, soil types *etc*., density of vegetation reflects the natural resource potential of watershed. This is assessed using satellite data, higher the density of vegetation, lower will be the priority.

Criteria adopted in prioritization are

1. **Waste lands**: These are described as degraded lands which can be brought under vegetative cover with reasonable effort and which is currently under utilized and deteriorating due to lack of appropriate soil and water

conservation measures. The waste land are identified based on the total variations observed on image due to lack of vegetative cover.

2. **Dry lands:** These are mainly single cropped areas, which are not supported by any irrigation source and are delineated using two season satellite data.
3. **Problematic soils:** The problematic soils include gullied lands, lands affected by salinity or alkalinity *etc.*, These are identified on satellite image through variation in tone, texture and association.
4. **Rainfall:** The areas which receives lower annual rainfall is given top priority when compared to the areas which receives normal annual rainfall. This was assessed based on the rainfall data collected for the preceding 10 years.
5. **Forest or vegetation cover:** The sustenance of forest cover is mainly related with soil properties, rainfall and other climatic conditions. The density of vegetation cover or forest is identified in the form of vegetation index through satellite data. If the vegetation index is low, then such areas are identified as high priority areas, which require immediate treatment.

Watershed characterization

The response of a particular watershed to different hydrological and geomorphological parameters. Though these are watershed specific and thereby unique, the characterization of a watershed provides an idea about its natural resource status. Watershed characterization thus involves measurement of parameters that influence the characterization behaviour of a watershed whereas, analysis aims at the critical study of these parameters to arrive at conclusions on watershed response and behaviour. Watershed characterization involves essentially inventory and mapping of natural resources.

I. Drainage and micro watershed delineation: The drainage lines are to be extracted from Survey of India toposheets and overlaid on satellite imagery (Fig. 28). Based on the presence of vegetation and water course some of the primary drains which were not observed in the toposheets are to be delineated. Water bodies of recent are also updated by comparing 3 season satellite images.

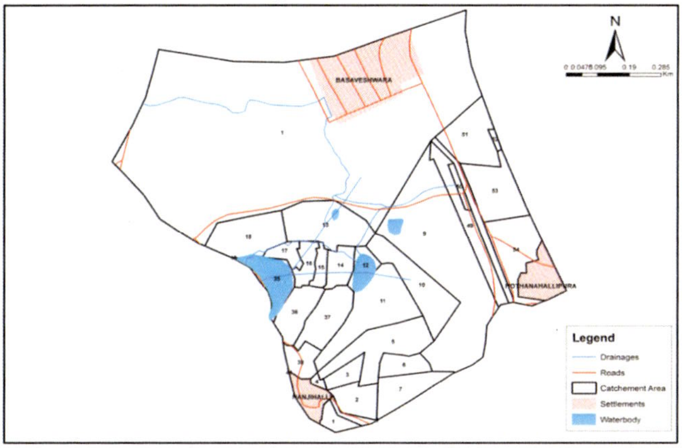

Fig. 28. Drainage map

II. Cadastral map: The village maps are procured from Land records and Survey settlement Department. The villages coming under one watershed are scanned. After scanning, maps are subjected for geo referencing using satellite images. Further mosaicing, digitization, attribute addition and boundary demarcation of these maps are done (Fig. 29).

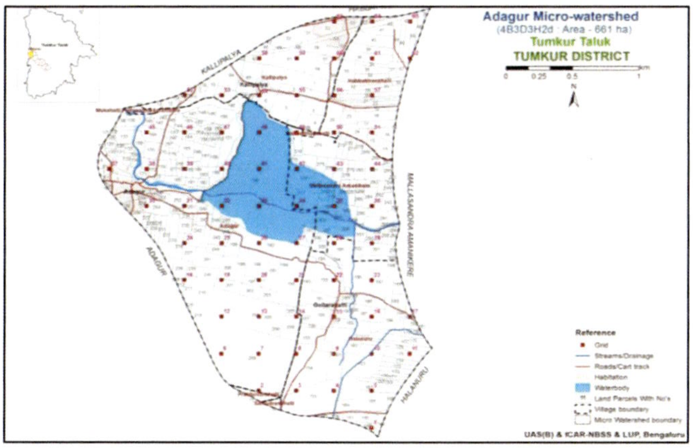

Fig. 29. Cadastral map

III. Slope: It is prepared using the contour information from Survey of India (SOI) toposheet (1:50000 scale). The major slope classes delineated are nearly level, very gently sloping, moderately sloping, strongly sloping and steep sloping (Fig. 30). The slope and aspect information is essential for delineating watershed boundary, planning soil and water conservation measures and for taking alternate land use practices.

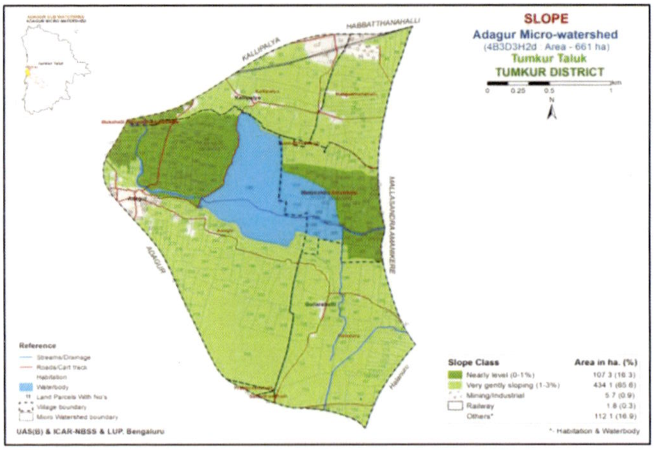

Fig. 30. Slope map

IV. Soil resource inventory and mapping: Soil are vital natural resources on whose proper use, depend the life-supporting system of an area and the socio-economic development of its people. For sustained utilization of soil resources, it is imperative to know the nature, characteristics and extent of distribution of different soils, their qualities and limitations. Soils are characterized, classified and evaluated for the suitability of land for a given use on the basis of a matching exercise between the growth and production requirement of the crop, and the physical or socio-economic conditions of the land and its users. The use of remote sensing data enables to map the soil boundaries accurately at relatively low cost and time. Thus making available the soil boundaries accurately at relatively low cost and time thus making available the soil resource potential and limitation of an area well in advance to take up suitable measures for improving the productivity. Soil map showing soil colour, texture, depth, erosion status, major limitations and land capability classes is prepared mainly by using the relationship between physiology and soils. Here major physiographic units are demarcated based on the contour information obtained from SOI toposheets on 1:50000 scale. The major physiographic units like hills and ridges, mounds, rolling lands, undulating lands, gently sloping lands, very gently sloping lands and valleys are further sub divided and based on the tonal variation on satellite image. Depending on the distribution and extent of these units, transects are selected, soil profiles are opened for analysing morphological and chemical properties. finally these physiographic units are converted in to soil units.

1. Soil Resource Inventory at Reconnaissance Scale Preparation of base map

A tracing film has to be overlaid on the toposheet covering the study area. Boundary of the taluk and important land features like rivers, tanks, roads, *etc.*, needs to be extracted. Thus a map having the above common land features will be used as a base map for preparing different thematic maps.

1. Land from analysis has to be made using Geology map.
2. A base map with permanent land features like roads, river and tanks along with major drainage lines prepared using survey of India toposheets is overlaid on geocoded FCC.
3. Based on image characteristics and contour information physiographic units such as hills and ridges, isolated hillock, mounds, rolling lands (10–15%), undulating lands (5–10%), gently sloping lands (1–3%) and valley (0–1%) are to be demarcated.
4. The number of physiographic units and their frequency of occurrence have to be listed.
5. Rapid traversing of the entire toposheet area has to be undertaken in order to check the physiographic delineation and correction of physiographic units needs to be attended wherever necessary.
6. The updated physiography map has to be scanned and digitized using software like Auto Cad, Arc GIS, Erads Imagine *etc.*
7. Considering the geographic distribution and frequency of occurrence of physiographic units in a toposheet, transects are delineated in such way that it should cut across at least three or more physiographic units.
8. In each physiographic units, profile are to be suited depending upon slope element or length of slope in order to establish relationship between physiography and soils.
9. Based on profile characteristics such as depth, colour, texture, gravelliness, structure, soil reaction, *etc.*, pedons were classified as per USDA soil taxonomy (Soil Survey Staff, 1996).
10. In this way, all the profiles in the project area are to be studied, classified and grouped into different series depending upon the above said characteristics.
11. For each physiographic unit, soil composition has to be worked out and the final composition so arrived at for a given physiographic units has to be checked at random (outside transect area) for the accuracy of soil composition.
12. Based on weighted average of different characteristics of all the profiles soil has to be grouped under each series.
13. Selection of typifying pedon has to be made for describing the pedon characteristics.

14. Horizon wise soil samples are to be collected from all master profiles of different series for laboratory analysis (pH, EC, OC, available N, available P_2O_5. CEC, bases saturation) using standard procedures.
15. The boundaries of physiographic units have to be merged depending on the soil analytical data and field observation and soil units boundaries will be arrived.
16. Based on final soil composition, soil map has to be finalized with legend showing classification depending upon the level at which one requires the mapping (Fig. 31).

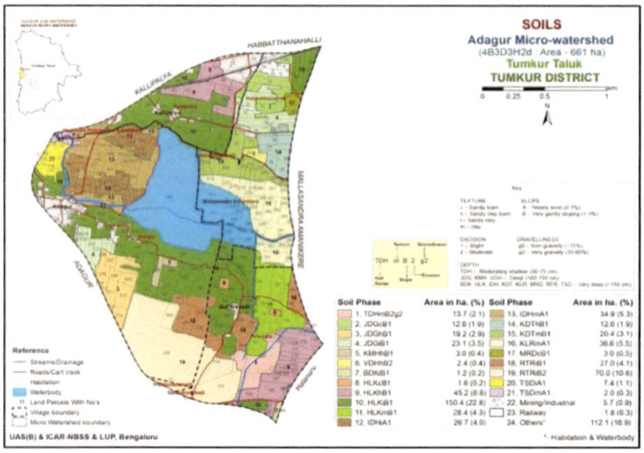

Fig. 31. Soil map

2. Detailed land resource inventory

The purpose of the resource inventory is to delineate similar areas, which respond or expected to respond similarly to a given level of management. This can be achieved in a village by the detailed study of all the sites (like slope of the land, erosion, drainage, occurrence of rock fragments *etc.*), and soil characteristic (like depth, texture, colour, structure, consistency, coarse fragments, porosity, soil reaction *etc.*), followed by grouping of similar areas based on soil-site characteristics in to homogenous (management) units and showing their extent and distribution on the villages cadastral map. The methodology followed in the field and laboratory is briefly indicated below (AIS & LUS, 1970, soil survey staff, 1993, 1996 and 2008).

Base maps

The detailed survey of the land resources occurring in the villages has to be carried out by using cadastral map as a base. The cadastral map shows field boundaries with their survey numbers, location of tanks, stream and other permanent features of the area

Apart from the cadastral map, remote sensing data products at the scale 1:8000 are to be used in conjunction with the cadastral map to identify the landforms and others surfaces features. The imageries helps in the identification and delineation of boundaries between hills, uplands and lowlands, water bodies, forest and vegetated areas, road habitation and other cultural of the area. Apart from the cadastral maps and imageries, toposheets of the area (1:50000 scale) are to be used for initial traversing, identification of geology and landform, drainage features, present land use and for the selection of transects at village level.

Field investigations

Preliminary traverse of the village has to be carried out with the help of cadastral map, satellite imagery and toposheets. During the traverse, based on geology, drainage pattern, surface feature, slope characteristics and land use, landforms and physiographic units have to be identified and initial legend needs to be prepared by studying soils in few selected places.

The field boundaries and survey numbers given on the cadastral sheet have to be located on ground by following permanent features like roads, cart tracks, nalas, streams, tanks, *etc*., and wherever changes are noticed they are to be incorporated on the cadastral map. After this, intensive traversing of each physiographic unit like hills, ridges, uplands, lowland/valley has to be carried out by involving all the soil survey party members. Based on the variability observed on the surface, transects are to be selected across the slope, covering all the landform units.

In the selected transect, profiles are to be located at closely spaced interval to take care of any change in the land feature like break in slope, erosion, gravel, stone *etc*. In the selected sites, profiles (vertical cut showing the soil layers from the surface to the rock) needs to be opened up to 200 cm or the depth limited by rock or hard substratum and studied in detail for all their morphological and physical characteristics. The soil and site characteristic are to be recorded for all profile sites on a standard proforma as per the guidelines given in USDA soil survey manual (soil survey staff 1993) Apart from the transect study profiles have to be studied at random, almost like in a grid pattern, between the transects.

Based on the soil-site characteristics, the soils will be grouped into different soil series (soil series is the most homogenous unit having similar horizons and properties and behaves similarly for a given level of management). Soil depth, texture, colour, amount and nature of gravel present, nature of substratum and horizon sequence are the major identifying characteristic of soil series occurring in the area.

Laboratory Characterisation

Soil sample have to be collected from representative profile for laboratory characterization. Surface samples are to be collected from farmers field (100 samples) for identifying status (major and micro nutrients). By linking the soil fertility data to the survey number through GIS, soil fertility maps can be generated.

Finalisation of Soil Maps

The area under each series has to be further divided into phases and their boundaries delineated on the cadastral map based on the variation observed in the texture of the surface soil, slope, erosion, presence of gravels, stoniness, *etc.* A phase is a subdivision of a soil series based mostly on the surface features that affect its use and management.

The delineated mapping units are shown on the map in the form of symbols. During the survey about 10 transacts representing different landforms occurring in the village have to be identified and profiles have to be opened for detailed soil and site characterization. In addition to the profile study, spot observations, in the form of mini pits and road cuts, terrace cuts *etc.*, are to be studied to prepare the soil map.

The soil map of the villages shows the distribution of mapping units occurring in the village. The series map shows the occurrence of soil series identified and mapped in the village.

The phase map, which shows the sub division of the series into the management units based on surface texture (Fig. 33), slope (Fig. 34), erosion (Fig. 35), gravels, rock out crops *etc.*, shows the distribution of phases mapped in the village. Each phase or mapping unit delineated on the map has similar soil and site characteristics. In other words, all the farms or survey numbers included in one phase will have similar management needs and they have to be treated accordingly.

Generation of Thematic Maps

Maps showing information about specific information for example soil pH, soil texture *etc.*, is known as Thematic map.

The soil-site characteristics are to be interpreted for identifying the constraints and potentials like soil depth (Fig. 32), erosion, gravelliness, salanity and land suitability maps for various crops by using GIS software have to be generated. The thematic maps thus prepared using the laboratory analytical data has to be analysed carefully for preparing the land capability map (Fig. 36) and land suitability map for major crops of the area (Figs. 37 & 38). The suitability map is prepared based on the matching exercise between the parameters required for taking up a crop and the actual parameters obtained from different thematic maps.

These different maps help in assessing the field condition before taking up any crop production. The farmer has to decide based on the suitability of their land to each crop and then take its cultivation to get maximum yield and economic returns. If a farmer takes up deep rooted crop/plantation in a shallow soil then crop establishment will be difficult and yield will be very low in this area. Hence based on the suitability of crop to each parcel of land, planning has to be made.

The thematic maps are prepared for each watershed in many of the watershed programmes of Government of India and Government of Karnataka. Under integrated Watershed Management Programme the thematic maps are effectively used by the implementing agency at various stages of planning, management and monitoring. Under Sujala III, a World Bank funded watershed project the different thematic maps are being developed in selected watersheds of different districts in Karnataka by various Agricultural Universities in colloboration with National Bureau of Soil Science and Land Use Planning and Indian Institute of Science, Bangalore. Karnataka State Remote Sensing Application Centre is the nodal agency for Karnataka to supply various maps to different user department.

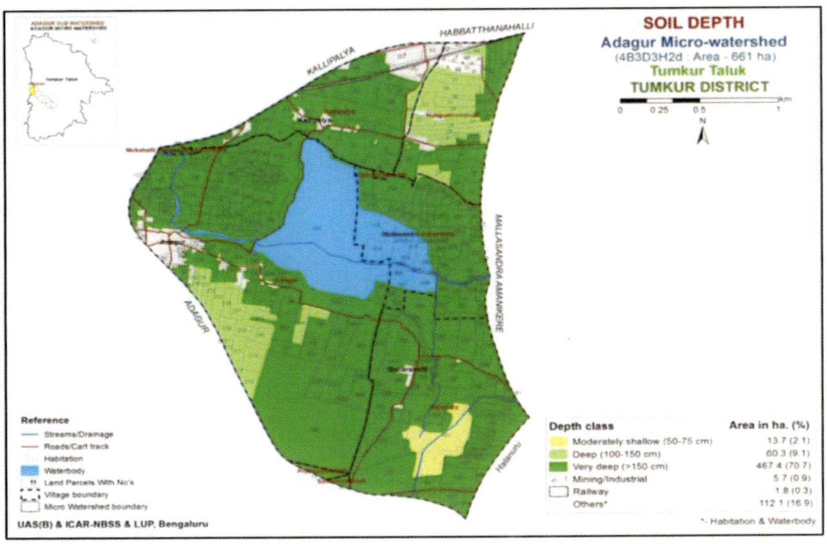

Fig. 32. Distribution of soil depth classes

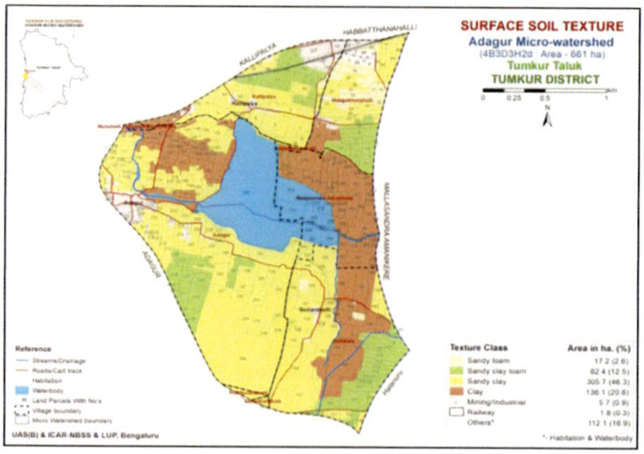

Fig. 33. Distribution of surface soil textural classes

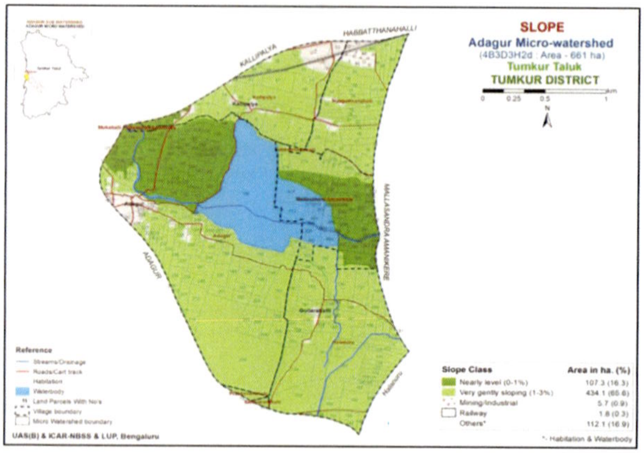

Fig. 34. Slope classes

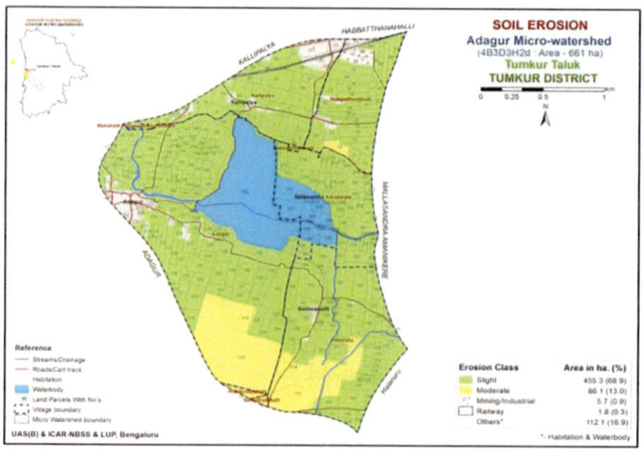

Fig. 35. Erosion classes

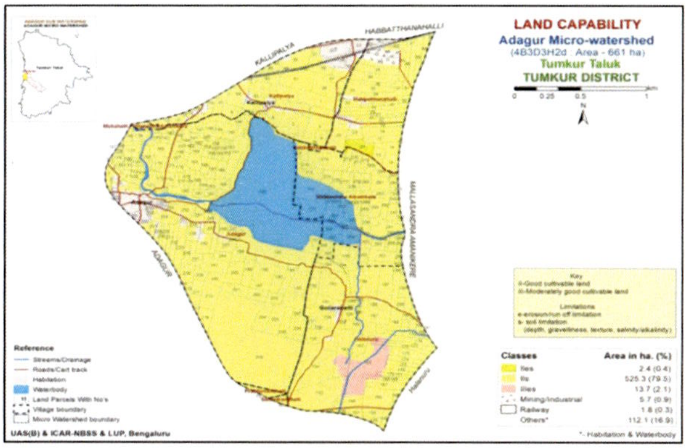

Fig. 36. Land Capability classes

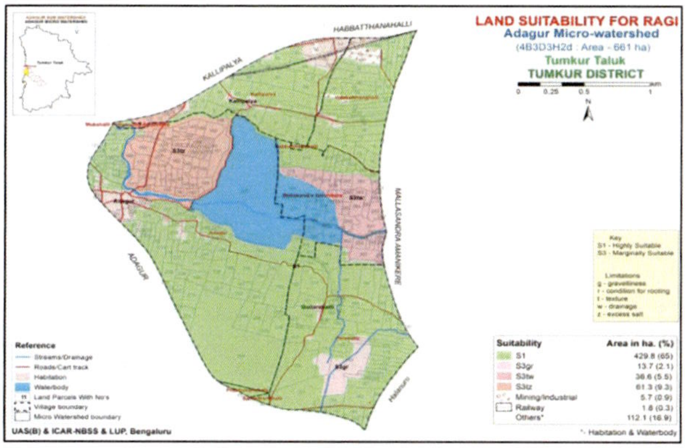

Fig. 37. Land suitability map for ragi

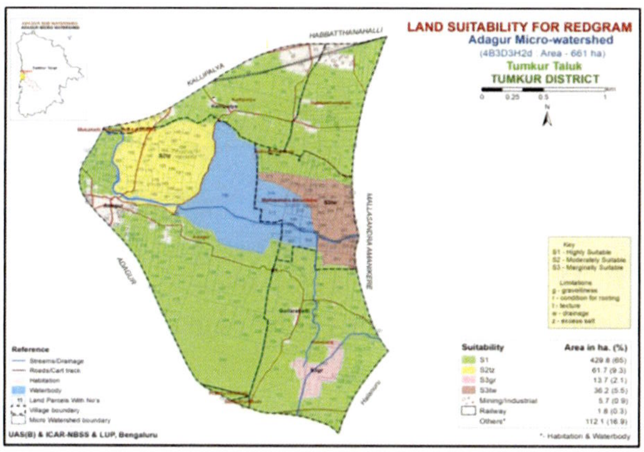

Fig. 38. Land suitability map for redgram

V. Ground water potential assessment: Remote sensing techniques play a vital role in preparing water resource development plan when used in the preliminary stages of the survey, the techniques helps in delineated potential areas of water bearing where further exploration work could be taken up. Thus, drastically reducing the cost and time involved in water exploration. Satellite imagery have been found to be particularly useful in identifying linear features such as lineaments, fractures, faults, folds, *etc*., which are ideal localizers of ground water sinks. Besides this information, the drainage pattern and drainage density, which have a bearing on the recharge conditions and permeability of the rocks. Satellite data thus offers a unique tool for extrapolating information on geology, geomorphology, drainage, land use and soils which are essential in identifying the possible location or zones of occurrence of groundwater. The groundwater potential map is prepared by obtaining the contour information from SOI toposheets, geomorphology, geology and tonal variation in the satellite imagery. The geomorphic units are then correlated with the ground water potential zones (Fig. 39).

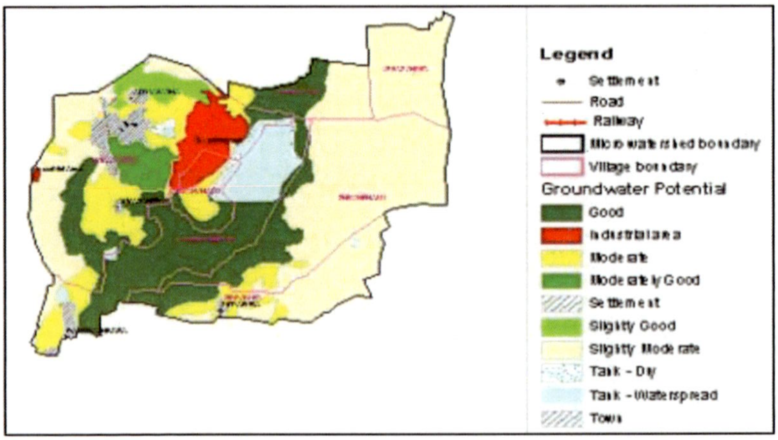

Fig. 39. Groundwater potential map

VI. Land use/Land cover analysis: The present land use/land cover map is prepared (based on the photo interpretation keys namely colour, tone, texture, association, size, shape *etc.*) using 3 season (*kharif*, summer and *rabi*) satellite data coupled with minimum field work, assessment of the extent and distribution of different land use land cover areas are made. The major categories like *Kharif*, double crop, *Rabi*, plantations, scrub land, gullied land, forest plantations, *etc*., are delineated (Fig. 40).

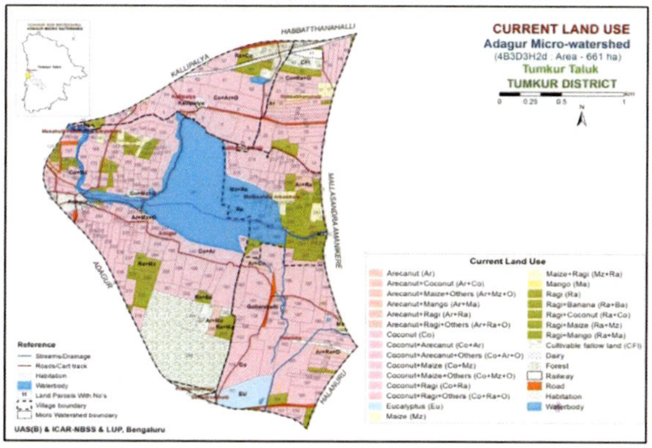

Fig. 40. Land use/land cover map

VII. Integrated resources development plan

A. Agriculture Action Plan: The Integrated resources development is a comprehensive action programme aimed at optimal utilization of resource potentials. Such an objective can be attained through development of land, water, vegetation and other resources of an area in sustainable manner so that the changes proposed to meet the needs of people are brought about without diminishing the potential for future. With the developments in satellite technology and data availability considerable progress has been made towards effective utilization of the available data for conservation, monitoring and management of natural resources. Integration of different thematic maps like Soil, Hydro geomorphology, Slope, Present land use land cover along with collateral data like Socio-economic data, Demand and supply, Meteorological data in GIS domain helps to generate database for planning quick remedial measures and permanent solutions to combat drought and other typical problems like drinking water, fodder and food grains through Agriculture resource development plan showing alternate land use practices like Intensive agriculture, Agro horticulture, Agroforestry, Fodder and fuel wood plantation *etc.*, (Fig. 41).

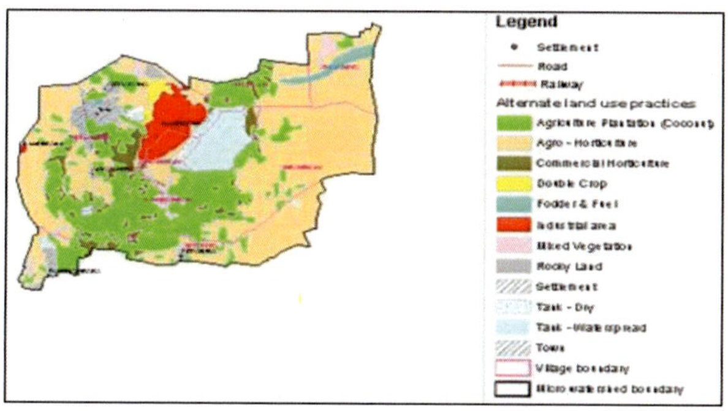

Fig. 41. Agriculture resource action plan

B. Water Resources Action Plan: The information on slope and soils along with drainage pattern is used to prepare Water resources development plan showing locations for water harvesting structures and soil conservation practices. These measures reduce the velocity of runoff, soil erosion and increase the infiltration. In addition desiltation of tanks, use of efficient irrigation systems and exploitation of groundwater in potential zones are also given priority after detailed field survey and study of hydro geomorphology map (Fig. 42).

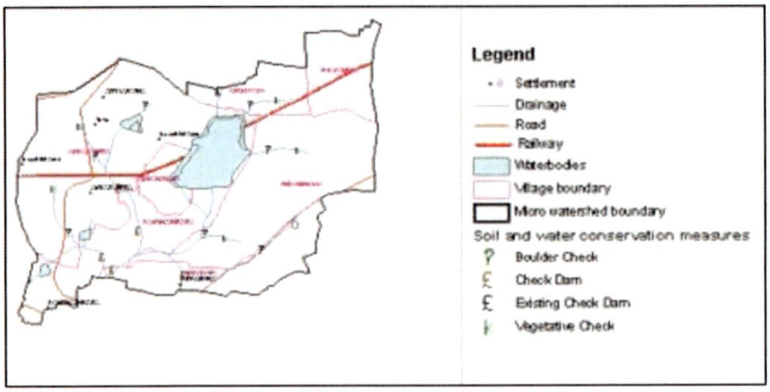

Fig. 42. Water resource action plan

Monitoring and evaluation of watersheds:

As the watershed development programme is an integrated and multidisciplinary approach implemented by spending considerable amount of money, monitoring of programme becomes an important component. It is necessary to assess and evaluate the long term effects and the impact of the developmental activities. Satellite Remote sensing data plays a vital role in this aspect by depicting the status of watershed before and after implementation of developmental programme, indicating the change detection and its impact on environment through comparison of images acquired for the same area. The monitoring and evaluation study thus would provide an insight into the impact of developmental activities in the watershed and also provide feedback for mid corrections (Fig. 43).

Fig. 43. Monitoring of watershed activities through GIS

Alternate Land Use System

Uncertain rainfall, poor soil conditions and low level of management has made annual cropping of field crops a non-remunerative enterprise in many pockets of dry lands. In some instances, cropping has been given up altogether and lands remain fallow and become wastelands overgrown with unwanted vegetation. To arrest this trend and to bring back the land under economically useful vegetation, alternate land use systems such as grasslands/pastures, agroforestry and horticulture are recommended. This is referred as alternate land use system. This is necessary because;

(a) Annual field crop production is nonviable and uneconomical in many years.

(b) Yield of field crops is low and fluctuates widely between years affecting stability and income.

(c) Continued use of the eroded and degraded lands led to the creation of wastelands.

(d) Alternate land use systems such as grasslands and tree culture are less risky, more productive and remunerative in these marginal lands. They will provide stability and sustainability.

The choice of an alternate land use system depends on the land capability. Most of the lands under dry farming tracts come under the land capability classes of III and above.

Land Capability Classification

Designated by the numbers 1 to 8. The numbers indicate progressively greater limitations and narrow choices for practical use. The classes are defined as follows:

Class 1: Soils have slight limitations that restrict their use.

Class 2: Soils have moderate limitations that restrict the choice of plants or that require moderate conservation practices.

Class 3: Soils have severe limitations that restrict the choice of plants or that require special conservation practices, or both.

Class 4: Soils have very severe limitations that restrict the choice of plants or that require very careful management, or both.

Class 5: Soils are subjected to little or no erosion but have other limitations, impractical to remove, that restrict their use mainly to pasture, rangeland, forestland or wildlife habitat.

Class 6: Soils have severe limitations that make them generally unsuitable for cultivation which restricts their use mainly to pasture, rangeland, forestland or wildlife habitat.

Class 7: Soils have very severe limitations that make them unsuitable for cultivation and that restricts their use mainly to grazing, forestland or wildlife habitat.

Class 8: Soils and miscellaneous areas have limitations that preclude commercial plant production which restricts their use to recreational purposes, wildlife habitat, watershed or aesthetic purpose.

Alternate land use system is applicable to all classes of land to generate assured income with minimum risk through efficient use of available resources. It aims at suitable farming system, matching the land capability classes. It is observed that Agri-horti system had less nutrient loss and more organic matter than either Agri or horti systems (Table 106).

Table 106: Nutrient loss in runoff under different land use system

Land use system	Nutrient loss in runoff (kg ha^{-1})			
	Org. matter	Nitrogen	Phosphorous	Potassium
Agri. System–Finger millet	21.8	2.9	0.27	4.4
Soybean	18.8	2.4	0.15	1.16
Agri-Horti Mango + Finger millet	16.7	1.96	0.11	1.15
Mango + Soybean	13.1	1.0	0.12	1.4
Sapota+ Finger millet	15.1	1.8	0.19	1.37
Sapota + Soybean	30.8	4.97	0.36	4.0
Horti- Pasture Mango + S. hamata	12.3	1.0	0.08	0.84
Sapota + S. hamata	0.05	0.01	0.001	0.01

Land use system	Nutrient loss in runoff (kg ha^{-1})			
	Org. matter	Nitrogen	Phosphorous	Potassium
Horti system Mango	20.2	2.5	0.1	2.3
Sapota	22.8	3.0	0.33	2.51

Land capability class Alternate land use recommended

Class II: Dry land horticulture

Class III & IV: Agro-forestry/leyfarming

Class V: Pastures/silvipasture/tree farming

Class VI: Range lands/wood lots

Pastures and grasslands

Forage crops play an important role in dry land economy. They help to promote livestock husbandry to improve and stabilize income. Forage grasses and legumes are best suited for marginal & sub marginal lands, sloppy lands, eroded and degraded lands for soil and moisture conservation and for reclamation of wastelands.

Forage crops for drylands include

Annual cereals : Sorghum, maize, pearl millet
Annual legumes : Cowpea, cluster beans (guar)
Perennial grasses : *Cenchrus ciliaris* (Anjan or Kolukkattai grass),
Cenchrus setigerus (black kolukkattai),
Cenchrus glaucus (blue buffel)
Dichanthium annulatum (marvel grass)
Chloris gayana (Rhodes grass)
Heteropogon contortus (spear grass)
Annual grass : *Pennisetum pedicellatum* (Deenanath grass).
Perennial legumes : *Stylosanthes hamata, S. scabra*
Leuceana leucocephala (subabul), berseem.

Forage crops can be introduced into the dry land farming system through

(a) Grasslands or pasture with perennial grasses and legumes for grazing by livestock, cutting and stall feeding, (cut and carry system.) and hay or silage making.

(b) Strip cropping with alternate strips of grasses/legumes and annual crops.

(c) Ley farming where in perennial forage crops are grown in rotation with annual crops in 4–5 year cycle Eg: *Stylosanthes hamata* (3 years) – sorghum (1 year) – castor (1 year).

Ley farming offers the following advantages

- Provision of fodder for cattle
- Low risk system
- Soil and moisture conservation
- Enrichment of soil fertility
- Prevention of soil compaction
- Control of perennial weeds

Silviculture: Refers to the raising of trees. When trees are introduced into farms along with field crops, it is known as agrisilviculture or agroforestry system. Trees provides many benefits to mankind. They play protective role by making available a variety of products for human consumption, for livestock and for industrial raw material needs. *Eg*: Fruits, nuts, fuel, fodder, timber, wood, wax, resin, *etc*. They also play a protective role through soil and moisture conservation, enrichment of soil fertility through nutrient recycling and protection of environment.

Different systems of tree culture

1. Agri-silviculture(Agroforestry): Trees and annual crops are raised in an intercropping system in the same field. Trees are planted at 5–8 m spacing and field crops are sown in the interspaces during rainy reason.

   Example: Leucaena + Sorghum/Pearl millet/Castor/Pulses
2. **Silvi-pasture:** Leguminous fodder trees are raised with fodder grasses and legumes as intercrops.

   Eg: Acacia + Cenchrus + Stylosanthes,
3. **Alley cropping or hedgerow intercropping**

   Annual field crops are grown in alleys formed by hedge rows of trees and shrubs. The trees or shrubs in hedge rows are cut back to short height (0.5–1.0 m) at sowing of annual crops with onset of rains and kept pruned during crop growing season to reduce shade effect and competition with field crops. The width of alley (space between hedges) is about 4–6 m.

   Example: Leucaena or Desmanthus as hedge row with sorghum, maize, pigeonpea, sunflower as intercrop. Alley cropping offers many benefits.

   Green fodder from hedge rows during dry season and food + dry fodder from annual crops.

- Off season rainfall is utilized by hedgerow trees or shrubs
- Hedge rows check runoff and erosion when formed along contour or across slope
- Loppings and prunings from hedgerows can be used as fodder, fuel wood or for mulching.

- Yield of crops raised in the alleys is improved due to better microclimate through reduction in temperature and wind speed, increase in humidity and reduction in evapotranspiration loss

Success in alley cropping depends on alley width and height of hedge rows.

Alley width of 5–6 m has been found to be effective. Low height of 45–50 cm is desirable. Usually one cutting of hedge row shrubs at the time of sowing of annual crops and subsequent prunings at monthly interval during cropping season are optimal. During dry season, cutting is done depending on fodder requirement.

4. **Timber-Fibre system (TIMFIB system):** It involves growing trees and perennial fibre crops together.

   Example: Leucaena + agave

   Choice of trees for drylands: Trees suitable for dry lands must have the following characters

- Multipurpose (fodder, fuel, timber, wood)
- Adaptable to wide variations in soil and climate
- Rapid growth
- Withstand severe pruning

5. **Agri-Horti systems:** Fruit trees *viz*, mango, sapota, guava, custard apple *etc*., can be established under low rainfall situation of dry regions. In the initial 5–7 years, location specific field crops can be grown in the inter spaces. Results of the trails at AICRPDA, Bangalore indicated that, Agri-horti system involving finger millet intercropped in mango alleys performed better over other systems in terms of growth and establishment of mango with less effect on the yield of the component crop (finger millet).
6. **Afforestation**: The upper reaches of the catchment in a given region are characterized by bare and rock out crops bearing witness to large scale denudation. This needs to be developed with *in-situ* soil conservation measures and appropriate choice of plant species to meet the needs of fuel, fodder, fruit, fibre, timber and other raw materials by the local community. The important plant species selected for developing forest lands through block plantation were *Acacia auriculiformis, Cassia siamea, Jambolana, Dalbergia sisso*, Tamarind, neem, Embilica, subabul, caliandra, *Albizia lebbeck,* custard apple, bamboo, pongamia *etc*. Khus and agave were introduced as filter strips on contour at a vertical interval of 1m. Catch pits were dug wherever soil was fairly deep in order to provide better soil moisture for the establishment and growing of plant species.

Integrated Farming Systems

Integrated Farming System is one where more than one agricultural activity is practiced in the same farm unit, the activities being interrelated and competes for the same set of available resources in the farm.

OR

Integrated Farming System is more or less stable combination of different crop, animal and other enterprises selected and operated for maintenance and utilization of all farm resources in order to obtain higher and sustainable production keeping in view the preference and goals of the farming family.

Why farming system?

- Soil degradation
- Waste and overuse of water
- Pollution of the environment
- Dependence of external inputs
- Loss of genetic diversity
- Loss of local control over agricultural production
- Global inequality

System

A System is an assembly of objects or things, which are interrelated and interdependent on each other for proper functioning to achieve a common objective Eg. Human body consists of digestive system, circulatory system and nervous system.

Farming System

Farming system is an integrated resource management strategy for obtaining economic and sustained crop and livestock production and preserving the resource bases with high environmental quality. It is a combination of cropping activity, animal husbandry and other related enterprises.

Farming system is a decision making unit comprising the farm household, cropping and livestock systems that transform land, capital (external inputs) and labour (including genetic resources and knowledge) into useful products that can be consumed or sold (Fresco and Westphal, 1988).

Classification of Farming Systems

1.	Climate	Tropical, Temperate, Mediterranean
2.	Stage of development	Shifting cultivation, Arable farming, Ranching.
3.	Components	Crop based, Dairy based, Poultry based, Fish based or Mixed farm.

4.	Mode of ownership	Peasant farming, Joint farming, Co-operative farming, State farming.
5.	Level of profitability	Subsistence, Commercial
6.	Irrigability	Wet, Dry, (garden/rain fed)
7.	Degree of mechanization	Mechanized, Semi-mechanized, Non-mechanized.

Farming System Components

Wetland	Garden land	Dry land
Cropping	Cropping	Cropping
Fishery	Milch cows	Goat
Poultry	Buffalo	Agro forestry
Pigeon	Bio gas	Horticulture
Goat	Spawn production	Tree
Duck	Mushroom	Pigeon
Pig	Homestead garden	Rabbit
Mushroom	Silviculture	Farm pond
Fodder	Sericulture	Fish

Integrated Farming System

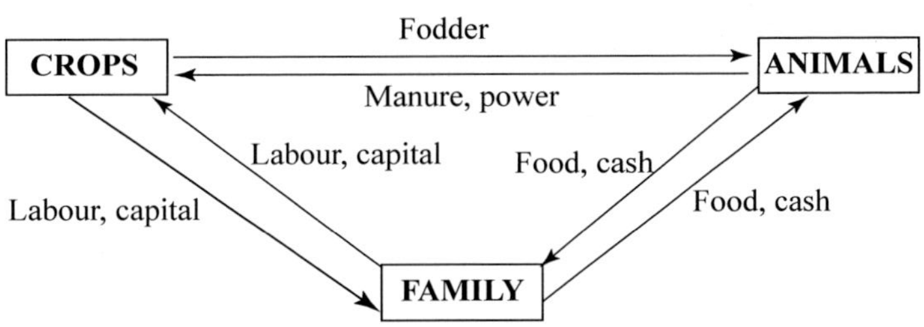

Scope and Need for Integrated Farming System

Agricultural development has undergone profound changes during the last two centuries. In western countries, machine power replaced animal power, fertilizers replaced organic manures. Dominance of fertilization and herbigation was noticed. Similarly in India, during the past four decades, crop improvement, improved production technologies, crop protection technologies were introduced. However, all these development activities were confined to agriculturally intensive 44 districts of our country. The improved technologies pose problems like

- High cost of inputs
- Fluctuation in market price of farm produce which does not commensurate with high cost of cultivation
- Environmental pollution *Eg*: Nitrate accumulation in Indus river.
- Soil degradation due to depletion of major secondary and micronutrients

- Low level of government support by way of giving subsidy to the farming community.
- Technologies developed for individual crop became economically not viable and sustainable. Hence the farmer has to link two or more allied enterprises to enhance or improve his income. In this context farming system play a vital role.
- Similarly drylands which constitute 99 million hectare and dryland farmers who are mostly marginal and small farmers (70% of farmer) face many problems.
- Due to uncertain rainfall the yield and income from the crop is uncertain
- The farm labourers are underemployed as the crop season is confined to four months only.
- Natural resource degradation *i.e.*, soil and nutrient loss due to water and wind erosion.

Under such situations the farming system approach forms the only solution to the poor, marginal and small farmers of India.

Basic principles in developing sustainable farming systems

- Sustainable agricultural system must be economically viable, dynamic and adoptable to changing needs and priority to renewable natural resources.
- Meet the needs of the present and future generations
- Prevent the soil degradation
- Integrated control mechanisms
- Avoid over exploitation of forests, pastures, *etc.*

Role of integrated farming system in sustainable agriculture

- India would need to produce additional food grains of 100 and 160 million tonnes by 2030 and 2050, respectively to feed its projected population.
- More than 85 million out of 105 million operational holdings in India are with less than 2 ha farm size, necessitating intensive cultivation involving the modern concept of time and space by adopting heavy dose of chemical fertilizers leading to accumulation of chemical residues in the soil (Swaminathan, 1990).
- As agriculture become intensive, soil toxicities and deficiencies of various kinds threaten the sustainable production.
- Yields start declining in high production area, if sustainability concepts are not integrated in different farming situations.

- One of the main objectives in IFS is recycling of produces/wastes among the components involved. The end products of such recycled materials are always used as organic manures for crop plants.
- A judicious mix of one or more enterprises with cropping complements each enterprise through effective recycling of waste/residues and in addition, when carefully chosen keeping in view the soil and environmental conditions.
- Some of the products are improved in their quality on recycling and thus the program provides double benefit to the farmer.
- Integrated farming system approaches are not only a reliable way of obtaining fairly high productivity with substantial fertilizer economy but also a concept of ecological soundness leading to sustainable agriculture.
- The need for system approach of nutrient supply to the system as a whole to increase the fertilizer use efficiency in economizing the costly mineral fertilizers by accounting residual effect of applied fertilizers and preceding crops was greatly emphasized.

Advantages of Integrated Framing System

1. Productivity

IFS provide an opportunity to increase economic yield per unit area per unit time by virtue of intensification of crop and allied enterprises. Time concept by crop intensification and space concept by building up of vertical dimension through crops and allied enterprises are the ways to increase productivity.

2. Profitability

The system as a whole provides opportunity to make use of the produce or waste material of one component as input on another component at the least cost. Thus by reduction of cost of production of individual components, the profitability per rupee invested is much enhanced.

3. Potentiality/Sustainability

Of late with an enthusiasm to produce more and more within the land area available to meet the requirement of additional population recorded at 2.2% every year, huge quantum of inorganic fertilizers, inorganic pesticides, fungicides, herbicides *etc.*, are dumped. Thus, there is likelihood of soil and environment becoming polluted and the productivity of the soil would automatically be reduced in the years to come. In IFS organic supplementation through effective utilization of by-products of linked components as manures is possible and thus it will certainly provide an opportunity to sustain the potentiality of the production base *viz.*, soil for much longer periods.

4. Balanced Food

In IFS, we link components of varied nature enabling to produce different *Source*s of nutrition *viz.*, protein, carbohydrate, fat, minerals, vitamins *etc.*, form the same unit area. This will provide an opportunity to solve the malnutrition problem that exists in the diet of the Indian farmers.

5. Pollution from environment

In the process of production in the crop-based activity, some of the organics are let as waste material and such materials when ignored as such will pollute the environment on decomposition. Similarly application of huge quantity of fertilizer, pesticide, fungicide and herbicide pollute the soil, water and environment to an alarming level. In the case of integrated farming system, waste materials are effectively utilised by linking appropriate components and thus utilizing the byproducts as organic manures. Similarly by adopting bio control measures will help in preventing pollution of soil, water and atmosphere. In areas where piggery is a major activity, dumping of pig dung gives out huge quantity of unwanted gas and pollutes the atmosphere in the longer run. In IFS, the pig dung is used in the biogas unit to produce bio energy for the utilization of lighting, lifting water from the underground, sources for running incubation unit of the poultry farm *etc.* Thus IFS absolutely helps in avoiding environmental pollution on longer run.

6. Recycling

IFS establish its stability due to effective recycling of produce or waste material of any one of the component as input on the other component linked in the program. Thus by way recycling their own material at the farm level, the farmer could able to reduce the cost of production which enables ultimately to increase the net income of the farm as a whole. Moreover it also helps in reducing the environmental pollution expected out of decomposition of organic residues of the farm activity.

7. Money round the year

Unlike conventional crop activity where the money is expected only at the time of disposal of the economic produce received after five to fifteen months depending upon the duration of the crop, the IFS provides flow of money to the farmer round the year by way of disposal of eggs, milk, edible mushroom, honey, cocoons of silkworm *etc.* This will help the resource poor farmer to get away from the clutches of usurious money lenders and from other financing agencies.

8. Adoption of New Technology

Most of the big farmers are fully aware of the impact of new technologies included in the package. But more than 80 per cent of the farmers who have been grouped under small and marginal category are not able to execute the advanced technology proposed for want of money. But IFS farmers, because of the linkage of Dairy/Mushroom/Sericulture/Fruit crops/vegetable crops/flower cultivation *etc.*, money received round the year by sale of produce from these components at weekly/fortnightly interval. Availability of cash round the year gives inducement to the small and marginal farmers to go for the adoption of technologies like fertilizer application, pesticide and herbicide application *etc.*, given in the package which otherwise is not possible under conventional farming due to paucity of funds.

9. Solve energy crisis

It is expected that the entire world is going to suffer for want of fossil fuel from 2030 AD. So it becomes inevitable to identify an alternative *Source* to solve our energy crisis within a span of 3 to 4 decades. In IFS by way of effective recycling technique the organic wastes available in the system can be utilized to generate biogas. Though this may not be a source for complete supplementation, at least to certain extent the energy crisis anticipated can be solved.

10. Solve fodder crisis

In IFS each and every cm^2 of land area is effectively utilized. Growing of perennial fodder legume trees in the borders and watercourses is a recommended practice in IFS. This practice not only helps in supplementing legume fodder but also enriches soil nutrient by fixing the atmospheric nitrogen. In the cropped land also IFS envisages intensification of cropping by including legume fodder like cowpea either as second tier or as third tier in the system. These practices will certainly relieve the crisis of non-availability of quality fodder to the animal component linked.

11. Solve fuel and timber crisis

The current production should be increased to eighteen folds in case of fuel wood and six folds in industrial wood. This could be possible to certain extent by afforestation program in the shrub jungles and sparse forest areas. In IFS by linking agro-forestry appropriately the production level of fuel-wood and industrial wood can be enhanced without detrimental effect on crop activity in the field level.

12. Avoid degradation of forests

There is a vast gap between the demand and production level as far as fuel wood and timber are concerned. This will naturally induce the users to encroach on the forest area nearby illegally to bridge the gap. Right now our forest area is less than (22%) the prescribed norms, (33%) to the geographical area available. When such encroachments are encouraged there is every possibility of our forest area going to be waste land in the years to come. Even in the forest area available at present more than 2/3rd is sparse forest. The statistics on soil erosion indicates that every year 5374 million tonnes of built up soil is eroded in our country. This is four folds higher than the prescribed norms *viz.*, 4 t ha^{-1} yr^{-1}. By linking agro-forestry in IFS, the degradation of forest area could be minimized to certain extent by supplementation of fuel and timber wood. By way of preserving the natural eco-system in the catchment areas, precious built up soil can also be preserved from erosion danger.

13. Employment Generation

Combining crop enterprises with livestock enterprises to take advantage to complementary and supplementary relationships between them, would increase the labour requirements tremendously and can help in solving the problem of under employment to a great extent. IFS provide enough scope to employ family labour round the year.

14. Improves literacy

The farmer, who adopts IFS by combining different components like fishery, sericulture, mushroom cultivation, apiary, spawn production, dairy, poultry, agri-horticulture, Agro-forestry-biogas production *etc.*, becomes an expert in each and every aspects of individual component on long-range adoption. This expertise will help the farmer to face any challenge in their activity.

15. Provides opportunity for Agri-oriented Industries

When once the produces of different components linked in IFS are increased to commercial level and if there is glut in the market, leads to the development of side industries for preserving the by products.

16. Increase input efficiency

The IFS provides enough scope to use the inputs on different components very effectively. This again leads to increased benefit cost ratio.

17. Increased standard of living of the farmer

When once provisions are made in the farm level to generate bio-energy, produce edible mushroom, fruits, eggs, milk, honey, vegetables *etc.*, for the family use of the farmer apart from commercial purposes through IFS, creates a sort feeling among the farmers that they are no way inferior to other professionals in the region. When they feel that the standard of living is on par with others and this will act as a booster tonic to continue agricultural profession without any reluctance, which exist at present among most of the farmers.

Steps involved in the preparation of a model Integrated Farming System to a specific situation

1. Assessment of available resources
2. Identifying the existing cropping system
3. Identifying component(s) to be integrated
4. Fixing the size of the individual components
5. Working out the requirement of the components
6. Modifying the existing cropping system to suit the requirements of the components integrated
7. Working out the economics of individual components and for the IF system as a whole
8. Identifying constraint and suggesting remedy measures/modifications for technical feasibility, economic viability and practical utility.

Enterprise linkage in IFS
Dryland ecosystem

Crop + Goat

Crop + Goat + Agroforestry

Crop + Goat + Agroforestry + Horticulture

Crop + Goat + Agroforestry + Horticulture + Farm pond

Crop + Goat + Buffalo + Agroforestry + Farm pond

Crop + Goat + Pigeon + Buffalo + Agroforestry + Farm pond

Crop + Goat + Rabbit

Models of Integrated Farming System for Drylands

The IFS under dryland farming can overcome risk, ensure optimum use of resources, improve the soil and moisture conservation and stabilize the income during period of drought.

For the sustainability of dryland agriculture, a holistic approach based on local resources of soil, climate and crop diversity would help in efficient use of natural resources apart from providing sustainability in food, income, nutrition, livelihood and economic security to the small and marginal farmers. The system should comprise food, fuel, fibre, fodder, fruit, livestock *etc.*, components in order to achieve sustainability.

Factors influencing the selection and size of components in Farming Systems

- Climatic conditions-Rainfall, temperature, solar radiation, wind, humidity, *etc.*

- Soil type
- Farmers preferences
- Size of the Farm
- Knowledge, skill and technology
- Storage, transport and marketing
- Resource mobilizing power of the farmer
- Credit facilities available
- Socio-economic condition

Module of IFS for Drylands

The dryland ecosystem is characterized by

1. Inadequate and uneven distribution of rainfall
2. Poor and marginal soils
3. Low cropping intensity
4. Limited crop diversification
5. Low value crops
6. Poor resource mobilizing power of farmers

The agriculture in drylands is seasonal. The cropping season is restricted to 4–5 months and people remain without employment for rest of the year. Diversification of cropping by integrating with components like livestock (Sheep/Goat rearing), silviculture, horticulture tree crops and pastures would improve the standard of living and employment opportunities of the dryland farmers (Table 107).

Integrated farming system is a boon to dry land farmers. When compatible components/allied activities/associated enterprises are suitably combined, farming in drylands becomes less risky and remunerative. **IFS on a watershed approach are the best way of alternative agriculture in dryland areas.** The following components are identified as appropriate to dryland situations.

Goat or Sheep rearing, Silviculture, Agroforestry, Farm Forestry, Horticultural tree crops.

Possible combinations are

Crop + Silvipasture + Goat/Sheep rearing

Crop + Silviculture + Horti. Fruit trees

Crop + Agro-forestry + Goat + Farm pond

Crop + Silvipasture + Goat + Pigeon + Farm pond

Crop + Silvipasture + Buffalo + Farm pond

***Eg.* Cropping + pigeon + goat + buffalo+ agroforestry + farm pond + border planting – 4.0 ha**

Farming system components	Area (ha)
CCS – Conventional cropping system with crop alone	1.0
FS_1 –IFS with crop + Pigeon + Goat + Agroforestry + Farm pond	1.0
FS_2 –IFS with crop + Pigeon + Buffalo+ Agroforestry + Farm pond	1.0
FS_3 –IFS with crop + Pigeon + Goat + Buffalo + Agroforestry + Farm pond	1.0
Border planting of *Glyricidia sepium* along field boundaries	
I. Cropping **Conventional cropping**	
Sole sorghum (CO 26) with recommended practices	1.0
Cropping system in IFS	
(i) Maize (F) + Cowpea (F) – Chickpea + Coriander	0.25
(ii) Sorghum (F) + Cowpea (F) – Chickpea + Coriander	0.25
(iii) Sorghum (G) + Cowpea (G)	0.20
(iv) Sunflower + Coriander	0.10
II. Agroforestry (*Acacia nilotica* + *Cenchrus ciliaris*) (Sorghum (F) + Cowpea (F))	0.10
III. Buffalo (2 milking & 1 calf) and (or) Goat (5+1 female: male)	0.05
IV. Pigeon (10 pairs)	0.01
V. Farm pond	0.04

Table 107: Productivity, Economics, Residue addition and Employment generation of the system

Farming systems	System productivity (kg ha^{-1})	Net return (₹ ha^{-1})	Employment generation (man days)	Residue addition (kg ha^{-1})
CCS	1270	1,167	30	1,688
FS_1	4723	9,304	113	3,855
FS_2	10994	22,670	141	10,383
FS_3	12387	21,818	163	11,583

(Esther shekinah, 1998–2002)

For the sustainability of dryland agriculture, a holistic approach based on biophysical resources would help in efficient use of natural resources apart from providing sustainability in food, income, nutrition, livelihood and economic security to the small and marginal farmers. The system should comprises of food, fuel, fiber, fodder, fruit, and livestock *etc.*, components in order to achieve sustainability (Table 108).

Table 108: Yield and economics of integrated farming system (1.0 ha-2008) at Chikkamaranahalli, Nelamangala taluk

S. No.	Cropping system	Area (ha)	Yield (kg $plot^{-1}$)		Returns (₹ha^{-1})		B:C ratio
			Grain or Pod	Straw or Haulm	Gross	Net	
1.	Groundnut + Castor (8:1)	0.07	55 24	299	1818	863	1.90
2.	Groundnut + Pigeonpea(8:2)	0.08	42 35	250	1846	676	1.58
3.	Pigeonpea + Field bean(8:2)	0.08	57 66	547	2795	1835	2.91
4.	Pigeonpea + Cowpea (1:1)	0.07	50 54	281	2341	1531	2.89
5.	Fodder maize + cowpea (3:1)– I crop Cowpea grain- II crop Fodder bajra-II crop	0.12 0.06 0.06		5500	4125	2745	2.99
6.	Samruddhi chilli (green)	0.03	181		1177	737	2.67
7.	Field bean	0.15	180	1583	4427	3027	3.16
8.	Finger millet + Pigeonpea (10:2) + mango	0.40	1150 106	3220	13841	9041	2.88
9.	Dairy				44000	22400	2.02
	Total	1.00			76770	42855	2.26

Source: Anon., 2008

Chapter 13

Strategies and Policies for Dryland Agriculture

Strategies to Improve the Productivity of Drylands

With renewed interest in rainfed agriculture during 1970s, both the national (AICRPDA and CRIDA) and international institutions (ICRISAT) have studied several aspects of rainwater management. Recommendations have been made after their through testing both at the research stations and on farmer's fields. Initially these technologies were accepted by the farmers because of the pursuasion by the Scientists and the incentives, but many of them are forgotten once the support is withdrawn. It is now time for us to examine whether the technologies themselves were weak or certain components were missing. Malfunctioning of the software may be due to a bug right from the beginning or it entered into the system at a later stage. It is also possible that we neglected certain critical inputs which became major constraints making the system defunct. Whatever the reason it should be identified and set right if the technology has to continue as tool for tackling the challenges of rainfed agriculture.

1. Proper agro-meteorological analysis of climate-soil-crop relationship for efficient crop planning and management
2. Basic approaches: Rainwater has remained an important input not only in crop production on drylands but also in providing drinking water for human beings and animals. Repeated experiments have proved that in vast areas of our watersheds, the rainfall is sufficient to at least 500 mm of rainwater in the surface one metre depth of soil which is sufficient to raise a reasonably good annual crop. The question here is not of how much rainwater is received in a watershed, but one of the quantum of rainwater stored in the soil profile and in usable storage structures. Three approaches have been traditionally adopted for management of rain water.
   (a) Effective *in-situ* water intake by facilitating infiltration and stretching infiltration opportunity time.

(b) Storage of surplus runoff in ponds and embankments for use during times of stress.

(c) Crop selection and management to efficiently use the conserved rainwater.

3. Crop improvement: Water availability is a major constraint in the production of dryland crop and it is expected to intensify with climate change. Better genetic tolerance to drought condition is a highly complex trait controlled by different genes at different plant development stages. In this background there is a need to screen genotypes for multiple resistances in different dryland crops with major emphasis on drought tolerance, short duration for climate extremities *etc.* Modern tools like biotechnology and nano science in evolving climate smart water productive genotypes.
4. Another important management practice is balanced fertilization including micro nutrients based on soil test results and distribution through subsidy chain in a single bundle along with seeds to dryland farmers can be up scaled.
5. Improvement of ecosystem of dryland through planting of Tree based oilseeds, glyricidia, pongamia and using the loppings for *ex–situ* incorporation in drylands for improving the soil health and fertility of drylands.
6. Establishment of nano-technology for soil and water conservation
7. Evaluation of soil moisture retainers *viz.,* hydrogels, zeolites *etc.,* and bio-irrigation technique as a drought proofing strategy.
8. Agro-techniques on socially acceptable cropping system (namely laser levelling, application of organic manures, modified method of sowing *i.e..* Aqua fertilizer drill and FIRB system of seed-bed preparation) basis are to be developed with a view to increase the FUE, WUE, efficient weed management, tillage system and crop-residue management Including INM.
9. To increase the intensity of cropping by developing appropriate intercropping and double cropping systems with emphasis on pulses and oilseeds.
10. Review of contingent crop planning for major aberrant weather conditions and improvement in the contingent crop planning for different zones mainly to mitigate drought effects.
11. In field conservation practices: The approach for a long time has been to construct contour or graded bunds in cultivated area with appropriate dimensions depending on soil type, land slope and rainfall distribution (Table 109) is required for optimization of resources.

Table 109. Choice of bunds for different soil and rainfall situations

Type of bunds	Rainfall region	Remarks
Contour bunds	Low rainfall region (<750 mm)	Suitable for medium soils of slope up to 6% not suitable for *Vertisols*.
Graded bunds	Medium to high rainfall regions (>750 mm)	Suitable for *Vertisols* when provided with up stream channel, in alfisols no upstream channel is needed.
Graded border strips	Medium rainfall Regions (750–1150 mm)	Suitable for deep *Alfisols* and vertic *Inceptisols*.

The question of cost, stability, loss of area and problems in maintenance of bunds have been the subject of debate for a long time. To reduce the cost of construction, smaller section graded bunds were preferred to large section contour bunds, especially in *Vertisols*. With the involvement of World Bank in our watershed activities vegetative bunds using vetiver took over earthen bunds. These bio-engineering structure acted as substitutes or supplement to the mechanical structures. Identification of alternative species was a subject for research for a long time and species like *Pennisetum hohenackeri* were found ideal. With all our knowledge on bunding we could not meet the requirements of the farmers especially the small and marginal farmers who wish to retain their boundary still meet the scientific criteria of conservation. An effort made in the early years of our research in operational research project at Rajanukunte is worth mentioning. As a deviation to the conventional contour or graded bunds the boundary bunds were retained and the existing water ways were renovated. One or two additional small section graded bunds were constructed and the land smoothing activity was introduced to provide a small gradient along upstream side of the bund leading to the water ways.

This approach needs further trials to identify the constraints in implementation. Another thinking that needs verification is the position of borrow pits in bund construction. Conventionally, the barrow pits are taken on the upstream side which will get filled up due to internal erosion over years. They also retain extra water for crops planted on the upstream side of the bund. But the problem with this arrangement is the prolonged time for filling up of the pits and the consequent water stagnation and yield loss. Instead, if the barrow pits are on the lower side of the bund, they could be leveled up using a wooden leveler to reduce the overall slope between bunds. Even if the the surface soil gets removed to an extent over a small distance, it will recover in a short period. This approach needs further testing in closely monitored watersheds.

12. Inter terrace management practices: Emphasis on "Location specific" research mainly to optimize the use of resources available for conservation and utilization of moisture under rainfed conditions. Large number of *in-situ* water conservation practices have been tested over years under different ago-ecological situations. Following are generally recommended under *Alfisols* and *Vertisol* watersheds, some are for increasing the infiltration and other are for smooth disposal of surplus water.
    (a) Land smoothing
    (b) Early tillage
    (c) Graded ridges
    (d) Furrow on grade
    (e) Contour cultivation
    (f) Inter cultivation
    (g) Increasing infiltration
    (h) Surface mulches
    (i) Vertical mulching
    (j) Compartment bunds
    (k) Scooping

    Through *in-situ* water conservation practices are simple to follow they are the ones least appreciated by several executives in watershed development department. The main argument is that there is not much for the departments staff to do and once the principles are explained, the farmers should adopt it by themselves. Several practices like 'fall ploughing', though appreciated by the farmers are not easily adopted as they lack the required power to take up cultivation on time. Hence, custom hiring centers at the village or hobli level with essential implements or farm power depending upon local soil, climate, crops grown and needs of the farmers can go a long way in adopting *in-situ* moisture conservation practices. Adoption of inter-terrace management practices alone can bring about 15–20% increase in productivity of dryland crops which can be harnessed by large scale demonstrations and Farmers Field Schools.
13. Soil health management: The rainfed areas are continuously exposed to various production constraints related to soil management. The major issues in *Alfisols* are crusting, moisture depletion, and decreased organic carbon content and soil fertility. These issues needs to be addressed through enhancing moisture holding capacity through alternate sources of organic manures, Crops and cropping system approach and Conservation agriculture.

14. Land Configuration: Inter terrace management practices are not necessarily for conservation of rainwater, at times they can be for smooth disposal of surplus water. When the time and intensity of rainfall are uncertain, a dry land farmer has to be ready with both the contingencies of surplus and deficiency. Hence, refinement of intercropping system with conservation furrow is needed for easy and large scale adoption.
15. Storage of runoff in the watersheds: Water harvesting and storage for re-use during times of scarcity is an accepted recommendation especially in low rainfall areas,. But, what is important is the estimation of the capacity required to store the runoff. Opinions vary on the proportion of runoff to be retained in the watershed and the portion that needs to flow out of the catchment. While some experts advocate 'no runoff' from the catchment, others feel that to maintain ecological balance and to serve the interest of downstream population, watersheds may retain 25–30% of the runoff and allow the remaining to flow through the streams. In this background, the guideline provided by Karnataka watershed development department to make provision for storing 25% of the runoff causing rains in ponds, nalabunds, check dams *etc*., needs to be re-examined.

    Stored water has multipurpose uses like recharge of ground water, protective irrigation to crops and drinking for animals and the storage capacity to be created depends on the distribution characteristics of rainfall. Each storage structure will fill up 3–4 times during the rainy season and therefore, the total capacity to be created could be about 30% of the total quantity of water to be stored.

    The main difficulty in estimating the extent of runoff is the non availability of the data on daily rainfall and rainfall intensity for each watershed/ catchment. In this direction, the efforts of the KSNDMC in Karnataka and similar organizations elsewhere to install telemetric rain gauges at panchayat land needs to be commended. Such data could be used for estimation of runoff in each watershed and storage structures planned accordingly.

    It is a common observation in watersheds that stored water in ponds is often not used for irrigations. It is not that farmers do not know the benefits of protective irrigation, but the reason is non availability of low cost and efficient water lifting devices. This single limitation has made the farm ponds defunct in several watersheds. Though some executive argue that the ponds are constructed solely for recharge of groundwater and least importance to be given to protective irrigation, this cannot satisfy small farmers as they have to lose a portion of their fertile land for pond digging and some more on its annual maintenance. If farm ponds have to be attractive to farmers, suitable lifting devices has to be identified and made available at affordable prices.

16. Pond sealing: Seepage losses from farm ponds could be very high especially in sandy soils. Past efforts to identify effective sealants have met only with partial success. Water is precious anywhere on earth but more so in low rainfall semi-arid regions. While we make all efforts to store runoff water, equal efforts are needed to identify cost effective sealants and efficient use of stored water for purposes like nursery raising growing high value horticultural crops, fish rearing and adoption of drip irrigation for fruit crops. Convincing data base on the economics of pond sealing and irrigation is yet to be developed.
17. Improvement, development and release of dual conditions high yielding varieties of food and vegetables crops for normal monsoon season and for dryland conditions. Availability of seeds of these varieties along with the suitable agronomic practices will certainly improve and stabilize the production and productivity of rainfed areas.
18. Proper adoption of alternate land use system including agro-forestry, agro-horticulture and fodder based cropping system in terms of diversification in cropping system required for saving water and for efficient use of water.
19. With the concerted efforts of AICRPDA, although number of technologies has been developed, the adoption is meager to the extent of 15–20% in case of slow moving soil and moisture conservation and crop management technologies with the exception of drought tolerant and high yielding varieties of dryland crops demanding invigorated extension activities.
20. Affordable power equipment which small farmers buy and maintain individually is essential to reduce drudgery and ensure timely agriculture operations in drylands.
21. Water sensitive planning encompassing management of runoff as a resource while protecting against the nuisance and damages of flooding through establishment of crop centric Best Management Practices (BMP), Soil Aquifer Treatment (SAT) for Sustainable Agriculture, small holder centric efficient water management practices through Application of communication and Information technology is required.
22. Commercial dryland agriculture under protected cultivation with harvested rainwater energy demand through harnessing solar energy in non-crop space including harvested water bodies to fulfill farm energy demand.

23. Post project maintenance: Much has been done in the area of watershed development both by the central and state Governments. Some states like Karnataka have created a separate department for this purpose. Commendable work is carried out during the period of implementation but the developed structures are not looked after by any agency after the conclusion of the project. Even good practices are forgotten. Therefore, it calls for identifying an agency which will be in charge of such assets and looks after its maintenance. Though on paper panchayats with user groups have to maintain the structures, it is not done in practice. Our suggestion to overcome this deficiency is to assign the maintenance by the individuals.
24. In spite of the availability of viable technologies for rainfed areas, most of them have not been adopted. The reason appears to be lack of convincing demonstrations. Such demonstration should focus on an integration of several aspects of water conservation and its use and should be laid out at the research stations, KVKs and on farmers fields.
25. Identification of the gaps in adoption and searching new technologies to fill such gaps is a continuous process. Each research station concerning dryland problems can adopt at least one watershed of about 500 ha and work in close association with development departments. The interventions here may not be necessarily the proven ones, but the ones that need to be vertified. The technologies to be adopted need not be rigid but the ones that can be modified based on previous experience. An action research approach to identify the gaps and solutions to the problems appears to be better suited to take the problems of rainfed agriculture.

Policy Issues in Rainfed Farming

For a number of reasons, the Government of India has heavy emphasis during the last thirty years on irrigated agriculture. But from the lessons learned from heavy reliance on the large dams, the government has begun to emphasize the importance of rainfed agriculture, beginning with the Sixth Five Year Plan (1980–81) to (1984–85). A comprehensive national watershed development programme was launched in 1984. However, due to a number of technological and socio-economic constraints, rainfed agriculture has not benefited sufficiently due to limited policy initiatives.

Singh (1990) made the following suggestions to strengthen the research and development of rainfed agriculture:

1. Undertake significant shift in investment from irrigated rainfed areas, with a major emphasis on afforestation and soil conservation projects organized on a watershed basis.

2. Investment in soil conservation practices including creation of infrastructure as well as water harvesting systems such as farm ponds has to be undertaken on a village-by village basis. This expenditure will only be a fraction of what the government already is spending on irrigation. Otherwise, without such government investment the high cost and the time delay in realizing immediate benefits will prohibit farmers from engaging in rainfed agricultural practices.
3. Weather aberrations continue to plague dryland farmers. Since seed is the primary input in the adoption of improved farm technology, seed banks must be established with government support in order to help farmers adopt contingent strategies, whenever the weather becomes aberrant.
4. Draft power is a serious content in dryland farming. Individual farmers cannot afford to purchase mechanical implements such as tractors. Therefore, a system of custom hiring in the villages would help farmers to complete their operations in a timely way at an affordable cost.
5. One of the principal factors explaining the non-adoption of improved crop production technologies is the element of risk involved in rainfed farming. Therefore, the availability of crop insurance for drought protection involving selected dryland crops would go a long way in instilling confidence in farmers and in stabilizing crop production.
6. Since farm holdings are excessively fragmented, further division of lands should be minimized by means of weaning away some population from agriculture. This can be done by encouraging the development of small scale, agro-based industries and services in the rainfed regions.
7. In addition to low productivity in rainfed areas there is also an acute scarcity of fodder. Silvi-pastoral systems on marginal lands should be encouraged through liberal financing, and fodder banks should be established to cope with scarcity during drought years (Ashok Dalawai, 2016).
8. Raising public investment in technology and infrastructure.
9. Rationalizing subsidies on agricultural inputs.
10. Covering all dry land crops under minimum support price schemes.
11. Making compulsory all dry land crops and livestock under insurance for all the households.
12. Inflow of institutional credit to dry land agriculture needs to be increased.
13. Considering dryland agriculture's role as a means of livelihood, enhancing its competitiveness by cutting the unit cost of production is critical for the survival of many smallholder farmers (Bantilan *et al.* 2003).

References

Anonymous, 1994, Integrated mission for sustainable development. Satellite remote sensing, Bulletin, DOS, Hyderabad.

Anonymous, 2006, Impact of shelterbelts in arid region of western Rajasthan. http://www.cazri.res.in/publications/KrishiKosh/108-(IMPACT.pdf.

Anonymous, 2008, Annual Report, All India Co-ordinated Research Project for Dryland Agriculture, Bangalore centre.

Anonymous, 2017, http://agrihelper.blogspot.com/2017/10/difference-between-dryland-farming-and.html.

Anonymous, 2019, Annual report, AICRPDA, UAS, G.KVK.

Anonymous, 2019,Double cropping using pond water

Anonymous, 2019, http://epgp.inflibnet.ac.in/epgpdata/uploads/epgp_content/ S000017GE/P001699/M020287/ET/1496119535Agro-climaticzones(2.pdf.

Anonymous, 2019, https://images.app.goo.gl/tq77XcYmKsrsAKxR8.

Anonymous, 2019,http://www.fao.org/3/T0115E0k.htm.

Anonymous, 2019,Performance of forage crops and chickpea under double cropping system.

Ainsworth, E.A., Beier, C., Calfapietra, C., Ceulemans, R., Durand-Tardif, M., Farquhar, G. D., Godbold, D. L., Hendrey, G.R., Hickler, T., Kaduk, J., Karnosky, D. F., Kimball, B. A., Körner, C., Koornneef, M., Lafarge, T., Leakey, A. D. B., Lewin, K. F., Long, S. P., Manderscheid, R., McNeil, D. L., Mies, T. A., Miglietta, F., Morgan, J. A., Nagy, J., Norby, R. J., Norton, R. M., Percy, K. E., Rogers, A., Soussana, J. F., Stitt, M., Wiegel, H. J. and White, J. W., 2008, Next generation of elevated [CO_2] experiments with crops: A critical investment for feeding the future world. Pl. Cell Environ.,31: 1317–1324.

Anonymous (1970), Soil Survey Manual, All India Soil Land Use Survey Organisation,

Aparna Bhide, 2019, Statistical validation of almanac predicted rainfall. M. Sc. Thesis submitted to the University of Agricultural Sciences, GKVK, Bengaluru

Babu, J.Y., Nayak, D.R. and Adhya, T.K., 2006, Potassium application reduces methane emission from a flooded field planted to rice. Biol. Fertil. Soils, 42:532–54.

Bellassen, V., Manlay, R.J., Chéry, J.P., Gitz, V., Touré, A., Bernoux, M. and Chotte, J.L., 2010, Multi-criteria spatialization of soil organic carbon sequestration potential from agricultural intensification in Senegal. Climatic Change, 98 (1–2): 213–243.

Bhatia, A., Sasmal, S., Jain, N., Pathak, H., Kumar, R. and Singh, A., 2010, Mitigating nitrous oxide emission from soil under conventional and no-tillage in wheat using nitrification inhibitors. Agriculture, Ecosystems & Environment, 136(3–4): 247–253.

Biradar, D. P., Shirahatti and Surakod, V. S., 2016, Dry Farming Research at Vijayapura: A Boon For Dry Land Farmers. In The Seminar on Holistic Management of Drought Stress Through Sustained and Futuristic Technologies organized by CoA, Vijayapura, UAS, Dharwad. Pp. 46–53.

Burney, J. A., Davis,S. J. and Lobell, D.B., 2010, Greenhouse gas mitigation by agricultural intensification. Proceedings of the National Academy of Sciences,107(26): 12052–12057.

Das, H. P., 1999, Management and mitigation of adverse effects of droughts phenomena in Sinha, D.K., Rahim, M. B. (eds.). Natural disasters-some issues and concerns, Natural Disasters Management Cell, Visva Bharati, Santiniketan, Calcutta, India, pp.87–103.

Derpsch, R., Friedrich, T., Kassam, A. and Hongwen Li, 2010, Current status of adoption of no-till farming in the world and some of its main benefits. Int. J. Agric. & Biol. Engg., 3(1): 1–25.

Dinnes, D.L., 2004, Assessment of practices to reduce Nitrogen and potassium non-point source pollution of Iowa's surface waters, Iowa Dept. of National resources, Des Moines, LA.

Directorate of Economics and Statistics, Ministry of Agriculture,2017

Dwivedi, R.S., Reddy, P.R., Sreenivas, K. and Ravishankar, G., 1999, A Remote sensing and GIS based integrated approach for sustainable development of land and water resources. Proc. ISRS Natn. Symp., Jan. 19–21, 1999, Bangalore: pp 352–360.

FAO, 1974,Irrigation and drainage paper, no. 25, Effective rainfall.

FAO, 1976, A Framework for Land Evaluation. FAO Soils Bulletin No. 32, FAO,Rome.

Giuseppe Di Girolamo and Lorenzo Barbanti, 2012,Treatment conditions and biochemical processes influencing seed priming effectiveness. Italian J. Agron., 7(25): 178–188.

Gopinath, K. A., Ravindra Chary, G., Srinivas Roa, Ch. and Narsimlu, B., 2016, Technologies for Climate Resilent Rainfed Agriculture. In The Seminar on Holistic Management of Drought Stress Through Sustained and Futuristic Technologies organized by CoA, Vijayapura, UAS, Dharwad. Pp. 1–11.

Guo, 1996, Straw Husbandry. Shanghai Science and Technology Press, Shanghai.

Hosamani, S. A., Patil. R. V. and Gulaganji, G. G., 1987, Effect of dry sowing and nitrogenlevels in sorghum genotypes. Sorghum Newsletter, 30–33.

http://ecoursesonline.iasri.res.in/Courses/Social%20and%20Farm%20Forestry/FRST201/Data%20Files/lec06.pdf

http://www.geographynotes.com/essay/climate-essay/classification-of-world-climate-essay-world-geography/3433

http://oar.icrisat.org/672/1/RA_00031.pdf

http://old.worldagroforestry.org/Units/Library/Books/Book%2006/html/8.9_trees_as_wind.htm?n=90

https://www.slideshare.net/chitranair16/remote-sensing-in-agriculture

https://www.slideshare.net/RambabuPalaka/characteristics-of-watershed

http://www.yourarticlelibrary.com/agrometeorology/top-14-indications-of-climate-aridity/88927

Iraimo, N and Munguamre, P, 2007, Rainwater harvesting technologies for small scale rainfed agriculture in arid and semi-arid areas. Department of rural Engineering, Faculty of Agronomy and Forestry Engineering, University Eduardo Mondlane.

IRRI, 2009, Every drop counts. Rice Today, 8(3): 16–19.

Itnal, C. J., 1981, Water conservation measures for increased production in black soils of Bijapur.

Khardiwar, M. S., Bhavin Ram, Shailendra Kumar and BharatSolanki, P., 2016, Study on effective monsoon onset and withdrawal in different taluka places of Buldhana district (M.S.). Journal of Environment, Ecology, Family and Urban Studies, 2(1): 1–8.

Krishnappa, A. M., 2006, Technical manual, Watershed development department, Bangalore.

Krishnappa, A. M., Arun Kumar, Y. S.,Gopala Reddy, T. and Nagaraju, 1994, Watershed Approach- A Boon for Dryland Agriculture: The Experience of Operational Research Project in Kabbalanala. Bangalore, India: University of Agricultural Sciences.

Krishnappa, A. M., Arunkumar, Y. S., Gopala, T. R. and Nagaraju, 1994, watershed approach- a boon for dryland agriculture: The experience of Operational Research Project in Karnataka, UAS, Bangalore.

Lal, R., 1998, Land use and soil management effects on soil organic matter dynamics on Alfisols in Western Nigeria. In: Lal, R, Kimble, J. M, Follett, R. F. and Stewart, B. A. Soil Processes and the Carbon Cycle. CRC Press LLC. 109–126.

Lalita Devi, 1992, Rainfall characteristics and water balance (A study of Uttar Pradesh). Concept publishing company, New Delhi-110059, 71–77.

Lehmann, J., Gaunt J. and Rondon, M., 2006: Bio-Char Sequestration In Terrestrial Ecosystems – A Review. Mitigation and Adaptation Strategies for Global Change11: 403–427. C Springer 2006, DOI: 10.1007/s11027-005-9006-5.

Lueders, T. and Friedrich, M. W., 2002, Effects of amendment with ferrihydrite and gypsum on the structure and activity of methanogenic populations in rice field soil. Applied and Environmental Microbiology, 68(5):2484–2494.

Maheswari, M., Sarkar, B., Vanaja, M., Srinivasa Rao, M., Srinivasa Rao, Ch., Venkateswarlu, B., Sikka, A.K., 2015, Climate resilient crop varieties for sustainable food production under aberrant weather conditions. Central Research Institute for Dryland Agriculture (ICAR), Hyderabad. P47.

Maheswari, M., Yadav, S.K., Shanker, A. K., Anil Kumar, M. and Venkateswarlu, B., 2012, Overview of Plant stresses: Mechanisms, Adaptations and Research Pursuit. In: Crop stress and its management: Perspectives and strategies. Eds. Venkateswarlu, B., Arun Shanker, K. and Maheswari, M. Springer Dordrecht, Heidelberg, London, New York. pp 1–18.

Manikandan, M., Thiyagarajan, G., Bhuvaneswari, J. and Prabhakaran, N. K., 2017, Wet and dry spell analysis for agricultural crop planning using markov chain probability model at Bhavanisagar. International Journal of Mathematics and Computer Applications Research, 7(1) :11–22.

Murthy, R.S., Hirekerur, L.R., Deshpande, S.B., Venkata Rao, B.V. and Shankaranarayana, H.S., 1982, Benchmark soils of India, NBSS & LUP, ICAR.

Natarajan, A. and Dipak Sarkar (2010) Field Guide for Soil Survey, NBSS&LUP, Nagpur.

Negi Mohita, http://www.yourarticlelibrary.com/soil/soil-erosion-paragraphs-on-soil-erosion-in-india/13895.

Pathak, P., 1984, Optimized land management system for Alfisols: research project report presented at the ICRISAT In-house Review. Patancheru, A.P. 502 324, India: International Crops Research Institute for the Semi-Arid Tropics.

Patil, N. D., Umrani, N. K. and Shende, S. A., 1968, Improved crop production technology for drought prone areas of Maharashtra. Technical Bulletin. Solapur, India: Dry farming Research Station. 96 pp.

Petersen, S. O., Regina, K., Pollinger, A., Rigler, E., Valli, L., Yamulki, S., Esala, M., Fabbri, C., Syvasalo, E. and Vinther, F. P., 2006, Nitrous oxide emissions from organic and conventional crop rotations in five European countries. Agriculture, Ecosystems & Environment,112(2): 200–206.

Piri, I.,Khanamani, A.,Shojaei, S. and Fathizad, H., 2016,Determination of the best Geostatistical method for climatic zoning in Iran.Applied Ecology and Environmental Research.15(1): 93–103.

Prabhakar, K.S., Ramesh, K.S. Lingaraju, Y. and Hegde, V.S, 1991, Characterization of watershed using IRS data- a case study covering Hesaraghatta catchment. Proc. workshop on Indian remote sensing satellite mission and its applications, Oct.25, 1991, Bangalore.

Prasanna Neginhal, M., 2010, Productive performance of intercrops in nipped castor (Ricinuscommunis L.) in Alfisols of Dryland. M. Sc. (Agri.) Thesis, Univ. Agric. Sci., Bangalore.

Raaga, 2019, Statistical analysis of rainfall of Bengaluru and Mysuru districts of Karnataka.M. Sc. Thesis submitted to the University of Agricultural Sciences, GKVK, Bengaluru.

Rajendra Prasad and Mertia, R. S., 2009, Tree windbreaks and their shelter benefits on farmland in arid region of Western Rajasthan. Journal of Soil and Water Conservation.,8 (4): 46–50.

Ram Mohan Rao, M. S., 1981, Moisture conservation for higher crop yields in black soils, Pages 48–53 in Panel discussion on soil and water conservation in red and black soils. Bangalore, India: University of Agricultural Sciences.

Ramachandrappa, B. K., Shankar, M. A., Dhanapal, G. N., Sathish, A., Jagadeesh, B. N., Indrakumar, N., Balakrishna Reddy, P. C., Thimmegowda, M. N., Maruthi, Sankar, G. R., Srinivasa Rao, Ch. and Murukanappa, 2013, Four decades of dryland agricultural researchfor Alfisols of Southern Karnataka (1971–2010), All India Coordinated ResearchProject for Dryland Agriculture, Directorate of Research, University of AgriculturalSciences, Bangalore.

Ramachandrappa, B. K., Sathish, A., Thimmegowda, M. N., Savitha, M. S., Srikanth Babu, P. N. and Anitha, M., 2014. Soil fertility assessment, mapping and fertilizer recommendation through geo-spatial approach. Technical Bulletin 01, AICRPDA, UAS, Bangalore, Karnataka, India. Pp 32.

Ramachandrappa, B. K., Thimmegowda, M. N., Balakrishna Reddy. P. C. Mariraju, H., Dhanapal, G. N., Shankar, M. A., Sathish, A., Jagadish. B. N. Indrakumar, N., Maruthi Sankar, G. R. Murukanappa and Srinivasarao, Ch. 2014. Thirty five years ofOperational Research Project, Bangalore 1976 to 2012: Achievements and Impacts.

Ramachandrappa, B. K., Shankar, M. A., Sathish, A., Alagundagi, S. C., Surakod, V. S., Thimmegowda, M. N., Shirahatti, M. S., Guled, M. B., Khadi, B. M., Jagadeesh, B. N. andSrinivasa Rao, Ch., 2014, Rainfed technologies for Karnataka. All India Coordinated Research Project for Dryland Agriculture, UAS, Bangalore and UAS, Dharwad. Pp. 1–105.

Rao, U.R, 1991, Remote sensing for sustainable development. Photo Nirvachak, J. Indian Society of remote sensing, 19(4): 217–235.

Reddy, K. S., Manoranjan Kumar., Rao, K. V., Maruthi. V., Reddy. B.M.K., Umesh, B., Ganesh Babu, R., Srinivasa Reddy, K., Vijayalakshmi and Venkateswarlu, B., 2012, Farm Ponds: A Climate Resilient Technology for Rainfed Agriculture: Planning, Design and Construction. Central Research Institute for Dryland Agriculture, Santoshnagar, Saidabad, Hyderabad-500059. Andhra Pradesh, India.

Sankar, M., Green, S. M., Mishra, P. K., Snoalv, J. T. C., Sharma, N. K., Karthikeyan, K., Somasundaram, J., Kadam, D. M., Dinesh, D., Suresh Kumar, Kasthuri Thilagam, V. andQuine, T. A., 2018, Nationwide soil erosion assessment in India using radioisotope tracers 137Cs and 210Pb: the need for fallout mapping. Current Science, 115(3):388–389.

Sathish, A., 1999, Remote sensing and Geographical Information Systems (GIS) applications in soil and agricultural resources management a case study of Bethamangala watershed. M.Sc. (Agri.) thesis submitted to UAS, Bangalore.

Shankar, M. A. and Premalatha, B. R., 2005, In- situ moisture conservation practices and rain water harvesting for stabilizing crop yields in drylands In: Compendium of lectures on strategies for efficient harvesting and recycling of rain water in relation to the improvement of water use efficiency in dryland crops. Pp: 17–27.

Shankar, M.A. and Premalatha, B.R., 2005, In-situ moisture conservation practices and rain water harvesting for stabilizing crop yields in dry lands In: Compendium of lectures on strategies for efficient harvesting and recycling of rain water in relation to the improvement of water use efficiency in dry land crops. pp. 17–27

Singh, H. P., Ramakrishna, Y. S., Sharma, K. L. and Venkateswarlu., 1999, Fifty years of Dryland Agricultural Research in India, Central Research Institute for Dryland Agriculture, Santhoshnagar, Hyderabad.

Singh, V. P., 1994, Elementary Hydrology.Prentice Hall of India Private Limited, 377–406.

Singlachar, M. A., Kulkarni, K. R., Poonacha, B. A, 2006, Technical Manual for Integrated Watershed Development. Watershed Development Department Government of Karnataka. pp. 745.

Smith, P., Martino, D., Cai, Z., Gwary, D., Janzen, H. H., Kumar, P., Mccarl, B., Ogle, S., O'mara, F., Rice, C., Scholes, R. J., Sirotenko, O., Howden, M., Mcallister, T., Pan, G., Romanenkov, V., Schneider, U., Towprayoon, S., Wattenbach, M. and Smith, J. U., 2008, Greenhouse gas mitigation in agriculture. Philosophical Transactions of the Royal Society B363:789–813.

Srinivasa Rao, Ch., Ravidra Chary, G., Mishra, P.K., Subbareddy, G., Maruthi Sankar, G. R. Venkateshwarlu, B. and Sikka, A.K. 2014, Rainfed Farming- A Compendium of improved technologies. All India Co-ordinated Research Project for Dryland Agriculture,Central Research Institute for Dryland Agriculture, ICAR, Hyderabad, pp.194.

Srinivasrao, Ch., Ravindra Charya, G., Rani, N. and Baviskar, V. S., 2016, Real time implementation of agricultural contigemcy plan to cope with weather aberrations in Indian agricultural Mausam, 67(1): 183–194.

Subramanian, K.S., Tenshia, V., Jayalakshmi, K. and Ramachandran, V., 2009, Role of arbuscular mycorrhizal fungus (Glomus intraradices) – (fungus aided) in zinc nutrition of maize. Journal of Agricultural Biotechnology and Sustainable Development. 1(1):029–038.

Takatori, F.G., Lippert, L. F. and Whiting, F. L., 1964, The effect of petroleum mulch and polythene films on soil temperature and plant growth. Proc. Am. Hort. Sci., 85:532.

Thyagaraj. C. R., Vital, K. P. R., Mayande, V. M. and Sharma, K. L., 1999, Tillage and soil management for higher productivity in drylands. P. 329–344. In Fifty Years of Dryland Agriculture, India.

Travis Lybbert and Daniel Sumner, 2010, Agricultural Technologies for climate change mitigation and adaptation in developing countries: Policy options for innovation and Technology Diffusion. Issue Brief No.6, ICTSDIPC Platform on Climate Change, Agriculture and Trade. Available at http://ictsd.org/downloads/2010/06/agricultural-technologies-for-climate-change-mitigation-and-adaptation-in-developing-countries_web.pdf.

Udhaya Nandhini, D. and Sakthinathan, B., 2017, Windbreaks and shelterbelts for soil conservation: A review. International Journal of Chemistry Studies., 1(2): 1–4.

Umbrani, N. K. and Bhoi, P. G., 1984, Effect of plant density on growth and yield of safflower under two rainfed situations. Indian J. Agron., 29(3): 262–266.

Venkataramana, G., 2001, A Strategy of sustainable development of typical watersheds using remote sensing and conventional techniques. Proc. of ICORG-2000,Vol. II. 2–5th Feb. 2001, Hyderabad: pp 184–190.

Venkateswarlu, B., Singh, A.K., Prasad, Y.G., Ravindra Chary, G., Srinivasa Rao Ch., Rao, K.V., Ramana, D.B.V. and Rao, V.U.M., 2011, District level contingency plans for weather aberration in India. Central Research Institute for Dryland Agriculture, Natural Resource Management Division. ICAR, Hyderabad. P.136.

Venkateswarlu, J., Vishnumurthy, T. and Padmanabhan, M. V., 1991, Contingency crop production strategies in rainfed areas under different weather conditions, Project Bulletin no. 5. CRIDA, Hyderabad.

Vijayalakshmi, K., 1989, Rainwater management in drylands. In: Technological advances indrylands, (Ed. S. P. Singh et. al.), CRIDA, Hyderabad, pp. 19–65.

Virmani, S. M. and Sivakumar, M. V. K., 1980,Climatic Classification: A Consultants' Meeting.

Vittal, K.P.R., Singh, H.P. Rao, K. V., Sharma, K. L., Victor , U. S., Chary, G. R., Sankar, G. R. M., Samra, J. S. and Singh, G. 2003. Guidelines on drought coping plans for rainfed production systems. All India Coordinated Research Project for Dry land Agriculture, Central Research Institute for Dry land Agriculture, Indian Council of Agricultural Research, Hyderabad. PP.39

Wassman R., Lantin R.S., Neue H. U., Buendia L.V., Corton T.M. and Lu Y., 2000, Characterization of methane emissions from rice fields in Asia. III. Mitigation options and future research needs. Nutrient Cycling in Agroecosystems,58: 23–36.

Wilhite, D. A., 1992, Drought, In: Encyclopedia of Earth System Science, Vol. 2. San Diego:Academic Press.

www.mapsofindia.com

Youkhana, A. and Idol, T., 2009, Tree pruning mulch increases soil C and N in a shaded coffee agro-ecosystem in Hawaii. Soil Biology & Biochemistry, 41:2527–2534.